PERGAMON INTERNATIONAL LIBRARY
of Science, Technology, Engineering and Social Studies
The 1000-volume original paperback library in aid of education, industrial training and the enjoyment of leisure
Publisher: Robert Maxwell, M.C.

ELEMENTS OF EXPERIMENTAL STRESS ANALYSIS

SI Edition

STRUCTURES AND SOLID BODY MECHANICS DIVISION

General Editor B. G. NEAL

Other Pergamon Titles of Interest

BRITVEC, S. J.	The Stability of Elastic Systems
CHEUNG, Y. K.	Finite Strip Method in Structural Analysis
DYM, C. L.	Introduction to the Theory of Shells
GIBSON, J. E.	Linear Elastic Theory of Thin Shells
HEARN, E. J.	Mechanics of Materials (in SI Units)
HERRMANN, G. & PERRONNE, N.	Dynamic Response of Structures
HEYMAN, J.	Beams and Framed Structures, 2nd Edition
JAEGER, L. G.	Cartesian Tensors in Engineering Science
JAEGER, L. G.	Elementary Theory of Elastic Plates
JOHNS, D. J.	Thermal Stress Analysis
LIVESLEY, R. K.	Matrix Methods of Structural Analysis, 2nd Edition
MALLOWS, D. F. & PICKERING, W. J.	Stress Analysis Problems in SI Units
PARKES, E. W.	Braced Frameworks—An Introduction to the Theory of Structures, 2nd Edition
ROZVANY, G. I. N.	Optimal Design of Flexural Systems—Beams, Grillages, Slabs, Plates and Shells
SAADA, A. S.	Elasticity: Theory and Applications
WARBURTON, G. B.	Dynamical Behaviour of Structures, 2nd Edition

ELEMENTS OF EXPERIMENTAL STRESS ANALYSIS

SI Edition

by

A. W. HENDRY

Ph.D., D.Sc., M.I.C.E., M.I.Struct.E., F.R.S.E.
Professor of Civil Engineering, University of Edinburgh

PERGAMON PRESS

OXFORD · NEW YORK · TORONTO
SYDNEY · PARIS · FRANKFURT

U.K.	Pergamon Press Ltd., Headington Hill Hall, Oxford OX3 0BW, England
U.S.A.	Pergamon Press Inc., Maxwell House, Fairview Park, Elmsford, New York 10523, U.S.A.
CANADA	Pergamon of Canada Ltd., 75 The East Mall, Toronto, Ontario, Canada
AUSTRALIA	Pergamon Press (Aust.) Pty. Ltd., 19a Boundary Street, Rushcutters Bay, N.S.W. 2011, Australia
FRANCE	Pergamon Press SARL, 24 rue des Ecoles, 75240 Paris, Cedex 05, France
WEST GERMANY	Pergamon Press GmbH, 6242 Kronberg-Taunus, Pferdstrasse 1, Frankfurt-am-Main, West Germany

First edition 1964

Reprinted 1968

SI edition 1977

Library of Congress Catalog Card No. 76-54580

Printed in Great Britain by Biddles Ltd., Guildford, Surrey

ISBN 0 08 021301 4 (Hardcover)

ISBN 0 08 021300 6 (Flexicover)

CONTENTS

PREFACE

THE object of experimental stress analysis is to deduce the stress conditions in a structural element subjected to some specified loading either by observation of the physical changes brought about in it by the applied loads or by measurements made on a model or analogue.

Ever since the principles of mechanics were first applied to engineering problems in the seventeenth century, recourse has been made to experiment as a means of elucidating the behaviour of structures. Early attempts at experimental analysis were inevitably crude by modern standards but, gradually, reliable methods and precise instruments were evolved which made possible detailed and accurate stress determination.

Sensitive strain measuring devices were available in the latter part of the nineteenth century, but they were generally too cumbersome for use on small elements and too delicate for use under field conditions. These limitations were eventually overcome but substantial progress in this branch of stress analysis was made only when electrical resistance strain gauges were introduced in the 1930's.

Photo-elasticity was known as a possible means of stress analysis from its discovery by Brewster in 1816, but again substantial progress did not take place until well into the twentieth century, in this case because suitable materials were not available until that time. Very few other methods for detailed stress measurement were developed before the 1930's but since then there has been very rapid progress and there are now techniques for the investigation of practically every kind of stress problem. Associated with these techniques is a range of equipment for applying and measuring loads and for measuring deflections, accelerations and other quantities which may be required for the assessment of stresses in particular cases. As previously suggested, the development of experimental stress analysis depends very much on developments in other fields of technology; for example, advances in strain gauge technique are closely tied to instrument technology in general and extension of the scope of photo-elasticity is almost entirely dependent on the emergence of new transparent plastics which happen to have suitable optical properties for this purpose. Those concerned with stress analysis must therefore look beyond the limits of their specialism in order to distinguish new possibilities and improved methods.

The purpose of this book is to describe the principles of the more important and widely used techniques and pieces of equipment and, by reference to a number of selected examples, to suggest appropriate applications of these in laboratory and field investigations. Although the examples used to illustrate the various methods of analysis are taken from the field of Civil Engineering, the book will be of use to all undergraduate and postgraduate students who require a basic knowledge of experimental stress analysis. It is also hoped that the book may be of use to practising engineers who may be concerned with experimental investigations in one way or another.

A book of this size covering such a wide field cannot possibly deal in great detail with any particular topic; a brief bibliography has therefore been given at the end of each chapter which should be sufficient to permit the reader to follow up the outline of principles given here in as great depth as may be required. Attention is directed in particular to the publications of the American Society for Experimental Stress Analysis; these include the very comprehensive *Handbook of Experimental Stress Analysis*, edited by Hetényi (Wiley, 1953) and the Proceedings of the Society which contain valuable papers on all aspects of the subject. Many papers relating to Experimental Stress Analysis are of course published in the journals of various scientific and professional institutions, in the technical press and by many research institutes. These may be traced by reference to publications such as Building Science Abstracts (H.M.S.O., London) and Applied Mechanics Reviews (American Society of Mechanical Engineers).

In conclusion, it may be added that although constant reference to books and periodicals is essential, real knowledge of experimental stress analysis can only be obtained by first hand experience in the laboratory.

ACKNOWLEDGEMENTS

Acknowledgements are due to the following for permission to make use of material from their publications in this text:

The Institution of Civil Engineers.

The Institution of Structural Engineers.

Rocha, M. M., Director, Laboratorio Nacional de Engenharia Civil, Lisbon.

Maihak A.G., Hamburg.

Sharples Engineering Co. (Bamber Bridge) Ltd., Preston.

I

MODELS, SCALE FACTORS AND MATERIALS

MODELS AND SCALE FACTORS

When experimental stress analysis techniques are applied in the design of structures or machines, it is usually necessary to employ reduced scale models in the investigations. Furthermore, it may be inconvenient or impracticable to make the model of the same material as the prototype or again, it may be desirable to use a model material of low elastic modulus in order to obtain easily measurable strains. It is thus essential that we should know the relationship between phenomena observed in a scale model and the corresponding effects in the full sized construction which it represents. It is, of course, necessary that this knowledge should be available at the outset in order that the model may be designed so as to represent correctly the behaviour of the prototype.

The problem of scales may be approached in two ways: firstly, by making use of formulae which represent the solution of a simpler problem of the same general type as that being investigated or, alternatively, by determining the required relationship by dimensional analysis.

As an example of the first approach, let us suppose that it is required to investigate the behaviour of a particular type of beam: suppose that the beam carries some specified loading, as shown in Fig. 1.1. Denoting the load on the prototype and on the model by appropriate subscripts and assuming geometrical similarity and homologous load systems, we can write:

	Prototype	*Model*
Load	W_p	W_m
Shear force	F_p	F_m
Shear stress (mean)	F_p/A_p	F_m/A_m
Bending moment	$k \, . \, W_p \, . \, L_p$	kW_mL_m
Bending stress	$M_p \, . \, y_p/I_p$	M_my_m/I_m
Deflection	$k_1 \, . \, \dfrac{W_pL_p^3}{E_pI_p}$	$k_1 \dfrac{W_mL_m^3}{E_mI_m}$

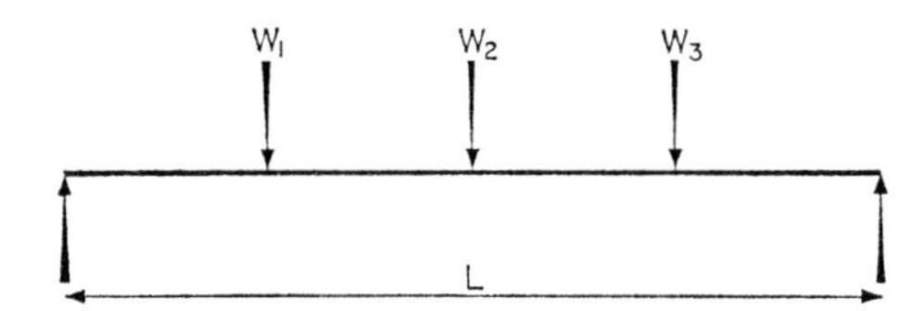

Fig. 1.1. Arbitrary loading on a simple beam.

Comparing the various quantities in the prototype and model we may derive the following ratios:

Load ratio = shear force ratio : $W_r = W_p/W_m$

Shear stress ratio: $$\frac{F_p}{F_m} \cdot \frac{A_m}{A_p} = \frac{W_r}{(L_r)^2}$$

Bending moment ratio: $$\frac{W_p}{W_m} \cdot \frac{L_p}{L_m} = W_r \, . \, L_r$$

Bending stress: $$\frac{W_p}{W_m} \cdot \frac{L_p}{L_m} \cdot \frac{y_p}{y_m} \cdot \frac{I_m}{I_p} = \frac{W_r}{(L_r)^2}$$

Deflection: $$\frac{W_p}{W_m} \cdot \left(\frac{L_p}{L_m}\right)^3 \cdot \frac{E_m}{E_p} \cdot \frac{I_m}{I_p} = \frac{W_r}{E_r \, . \, L_r}$$

In this way it is possible to determine the relationship between corresponding quantities in the model and prototype and thus to interpret model observations in terms of full size. Examination of the relationships for bending stress and deflection show that it is not necessary for the cross section of the model member to be geometrically similar to that of the prototype, provided that the ratio of the moments of inertia is introduced. The final result quoted above of course assumes geometrical similarity throughout. This method can be applied to any problem for which a type solution is known.

The alternative approach by dimensional analysis possesses greater generality as knowledge is required only of what variables control the behaviour of the system. It is beyond the scope of this book to deal with the subject of dimensional analysis and only an indication of the method of approach will be attempted; a full treatment of the method will be found in reference 1.

If we consider the same problem as discussed above, it is easily seen that the stress in the beam depends only on the loading on it and on its dimensions, i.e.

$$\sigma \propto W^a L^b \tag{1.1}$$

Now the dimensions of the various terms in (1.1) expressed in fundamental units of force F and length L are:

$$\sigma \quad [FL^{-2}]; \qquad W \quad [F]; \qquad L \quad [L]$$

Thus:

$$[FL^{-2}] = [F]^a[L]^b \tag{1.2}$$

Since the equation must be dimensionally homogeneous we may equate indices of like dimensions on the two sides, so that:

$$a = 1 \qquad b = -2$$

This gives:

$$\sigma \propto \frac{W}{L^2} \tag{1.3}$$

Comparing the stress in the model with the stress at a corresponding point in the prototype:

$$\frac{\sigma_m}{\sigma_p} = \frac{W_m}{W_p} \bigg/ \left(\frac{L_m}{L_p}\right)^2 = \frac{W_r}{(L_r)^2} \tag{1.4}$$

which is the same result as obtained by the alternative method.

The deflection scale ratio can be found in a similar manner as follows:

$$y \propto W^c L^d E^e \tag{1.5}$$

Deflections are proportional to strains and therefore to stresses; these are in turn proportional to the applied load. Thus in (1.5) $c = 1$. Expressing this equation in terms of dimensions we obtain:

$$L = (F)(L)^d(FL^{-2})^e$$
$$= F^{1+e}L^{d-2e}$$

and thus:

$$1 + e = 0 \qquad d - 2e = 1$$

whence:

$$e = -1 \qquad d = -1$$

and

$$y \propto \frac{W}{EL}$$

The deflection scale ratio is therefore:

$$\frac{y_m}{y_p} = \frac{W_m}{W_p} \cdot \frac{E_p}{E_m} \cdot \frac{L_p}{L_m} = \frac{W_r}{E_r L_r} \tag{1.6}$$

MATERIALS FOR EXPERIMENTAL STRESS ANALYSIS

The material adopted for the construction of a model depends on the nature of the structure or element to be represented. Obviously, a first requirement is that the stress–strain relationship of the prototype material should be correctly represented: if this is elastic then the model material must also be elastic although, as we have seen in the preceding paragraph, the elastic moduli need not be identical. The model material must of course be sufficiently strong to withstand the imposed stress. A second requirement is that the material should be capable of being cast or machined to the shape of the prototype with reasonable ease.

Table 1.1 shows Young's modulus and Poisson's ratio for a variety of materials which have been found valuable for experimental stress analysis.

Table 1.1

Properties of Materials

Material	E N/mm²	*Poisson's ratio*	*Remarks*
Metals			
Steel	200×10^3	0·30	
Brass (70–30)	107×10^3	0·33	
Duralumin	67×10^3	0·33	
Plaster and Cement			
Plaster of Paris	$2·2 \times 10^3$	0·15	
Portland cement mortar	13×10^3	0·15	Water–cement ratio approx. 1·25:1
Plastics and Rubber			
Methyl methacrylate	$3·3 \times 10^3$	0·38	Perspex
Phenol formaldehyde	$2·0 \times 10^3$	0·42	Catalin
Polyethylene	$0·35 \times 10^3$	0·50	Alkathene
Polyester resin (15°C)	$3·0 \times 10^3$	0·40	Araldite MY753
,, ,, (80°C)	$0·01 \times 10^3$	0·49	,,
Rubber	0·7–0·14	0·50	

Metals have well defined elastic properties, but their comparatively great rigidity and the difficulty of forming them to complex shapes limits their use for model analysis. The high elastic moduli of metals means that large loads may be necessary to produce measurable strains requiring substantial and expensive loading equipment. There are, of course, situations in which the high accuracy to which metals can be machined and their definite elastic properties may outweigh the difficulties. In general, however, it will be found convenient to use a plastic or other non-metallic material. The properties of these materials are very variable depending on the precise composition and in many cases on their age; the values quoted in Table 1.1 are thus to be regarded as typical. Furthermore, all these materials are subject to mechanical creep, i.e. for a given stress, the resulting strain increases with time. This factor is most important in experimental stress analysis and must be allowed for in the experimental technique. Fortunately, most of the materials employed for models are almost perfectly visco-elastic, that is the stress–strain relation at any instant after loading is linear although the modulus of elasticity varies with time. This behaviour is illustrated by the stress–strain–time curves shown in Figs. 1.2 and 1.3 which are for the material Alkathene.

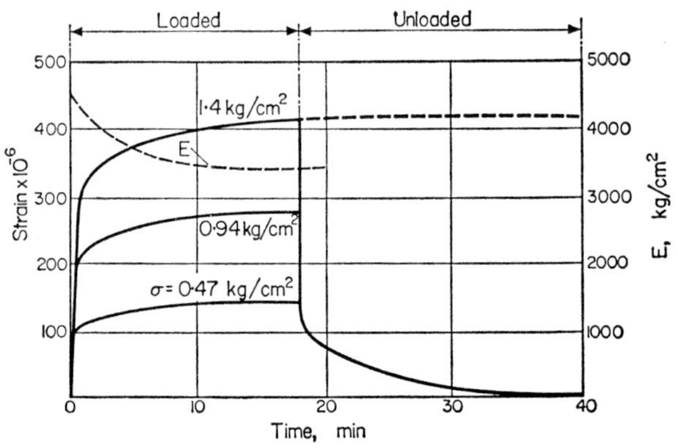

Fig. 1.2. Creep tests on Alkathene prisms (after Rocha[2]*).*

Difficulties arising from creep may be overcome either by deferring measurement until the increase of strain with time is very small or by taking readings at a particular time interval after loading; calibration tests to determine elastic moduli must, of course, be carried out at the same length of time after loading as in the main experiment. It is also important to see that there is no appreciable creep during the time necessary to take readings on the model. A method for obtaining a considerable number of readings from a model of visco-elastic

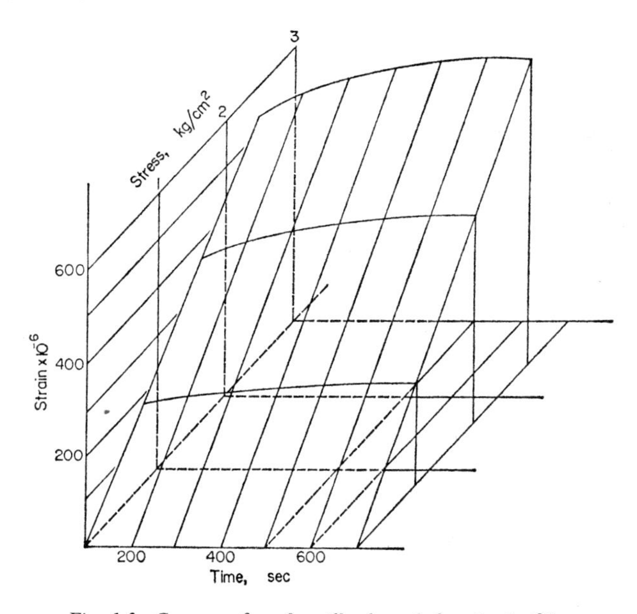

Fig. 1.3. Creep surface for Alkathene (after Rocha[2]*).*

material has been described by Rocha[2]; this may be understood by reference to Fig. 1.4 which shows the strain produced by repeated loading and unloading of such a material. It will be seen that after two or three equal cycles of loading and unloading, the difference in strain observed between loading and unloading becomes constant. In other words the material behaves as if it were elastic; the effective elastic modulus is based on the strain difference between the beginning of loading and the beginning of the following unloading.

As may be seen from Table 1.1, all the plastics have high values of Poisson's ratio. In models of structures in which deformations are controlled predominantly by one of the

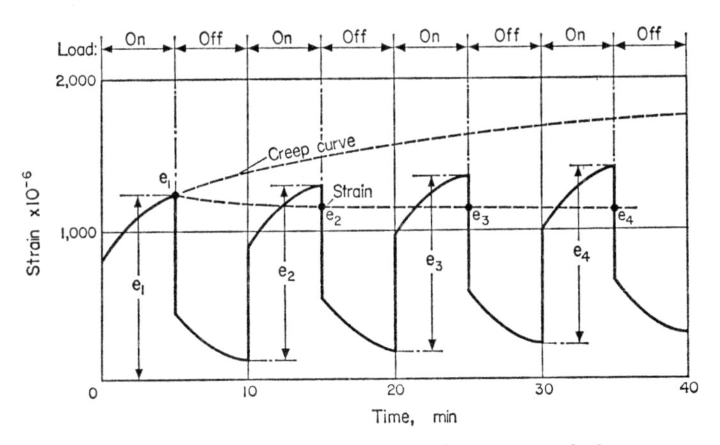

Fig. 1.4. Repeated loading of a visco-elastic material showing constant strain increment after first loading cycle (after Rocha[2]).

elastic moduli, as in a beam problem where the elementary Bernoulli–Euler bending theory is applicable, the question of relations between the elastic moduli does not arise. In cases involving complex stress systems the stress distribution may be appreciably influenced by the value of Poisson's ratio[3,4]. It has been shown theoretically, and confirmed by experiment, that the stresses in elastic plates are independent of the ratio of the elastic contents provided that there are no holes. If there are holes this statement is only true if the forces applied to the boundary of any one hole form a system in equilibrium or reduce to a couple. If there is a resultant load on the boundary of a hole, a correction is necessary if the results are to be transferred from a plate of one material to a plate of a different material. Fortunately, the correction is usually small and the error involved in transferring from one material to another is rarely more than a few per cent.

Of the various materials mentioned, each has its particular characteristics, and selection must depend on the nature of the problem. As previously mentioned, the use of metals is restricted by their great rigidity and the difficulty of shaping models from them. It is quite easy to make up models of rigid frame structures from metal strips, but it is difficult to get a sufficiently wide range of sections for practical purposes; variations of moment of inertia in a particular member of course adds considerably to the difficulties. If the behaviour of the material above the elastic range is being studied, it may be impossible to find any other material exhibiting the required properties and in such a case the difficulties associated with the use of metals have to be overcome.

Plaster of Paris is very useful for models of concrete structures such as dams [2] and has in recent years been used for representing reinforced concrete [5, 8].

The advantages of plaster include quick setting times and the possibility of casting thin sections of adequate strength. It must be realised, however, that the tensile strength of a plaster mix is likely to be proportionally higher than that of concrete and thus care is necessary in interpreting the results of plaster model tests in terms of concrete. By varying the water/plaster ratio a material can be prepared having a value of E, when set, between approximately 3930 and 393 N/mm^2. This is useful in the analysis of dams when the elastic modulus of the underlying rock may differ considerably from that of concrete. The admixture of diatomite may be necessary to secure workability of drier mixes and the tensile strength can be reduced by the addition of sand, e.g. a mix consisting of equal parts of plaster and sand produces a material suitable for models of concrete structures.

Plastics are extremely valuable for model construction [2, 4] both in sheet and cast forms. The most widely used sheet material for this purpose is Perspex, which is an acrylic resin;

it is easily cut, machined and jointed and is commercially available in a range of thicknesses. For casting, there are various epoxy resins, such as Araldite MY753, and polyethylene of which Alkathene is a well known example. Araldite MY753 is cast at room temperature but considerable heat is generated as polymerisation of the material takes place. This limits the thickness of castings owing to the possibility of cracking, particularly if the casting is of complicated shape. Polyethylene has to be cast at over 130°C and thus requires metal moulds. In addition, high shrinkage takes place necessitating great care in the design of moulds and in casting technique.

Materials for photo-elastic analysis have to possess suitable optical properties as well as adequate strength, absence of creep and good machining characteristics. Suitable materials for this purpose will be mentioned in Chapter VII.

BIBLIOGRAPHY

1. H. L. Langhaar, *Dimensional Analysis and Theory of Models.* Wiley 1951.
2. M. Rocha, *Model Study of Structures in Portugal*, Lab. Naç. de Eng. Civ., Lisbon, Pub. No. 84 (English trans. available as NRC TT-970, Nat. Res. Council of Can. Div. of Bldg. Res., 1961).
3. A. W. Hendry, *An Introduction to Photo-elastic Analysis*, Ch. 1, pp. 16–18. Blackie, Glasgow, 1948.
4. M. Clutterbuck, The dependence of stress distribution on elastic constants, *Brit. J. App. Phys.*, **9** (8), 323–8 (1958).
5. G. Brock, Direct models as an aid to R.C. design, *Engineering*, **187** (4857), 468–70 (1959).
6. R. G. Smith and C. O. Orangun, Gypsum plaster models of unbonded prestressed concrete beams, *Civ. Eng. Pub. Wks. Rev.*, **56** (660), 906–9; **56** (661), 1061–3 (1961).
7. E. Fumagalli, Suitable materials for static and dynamic tests on model concrete dams, Proc. 5th Cong. on Large Dams, Paris, Vol. 4, pp. 1039–74, 1955.
8. E. Fumagalli, The use of models of reinforced concrete structures, *Mag. of Conc. Res.*, **12** (35), 63–72 (1960).
9. R. Meadows, Deflection tests of plastic models, *Proc. Soc. Stress Anal.*, **8** (1), 117–128 (1950).
10. Effect of quantity of hardener on elastic and photoelastic properties of Araldite casting resin **B** at 20°C and 150°C, *Schweiz. Archiv.*, **8** (1955).

II

LOAD APPLICATION AND MEASUREMENT

ONE of the most important considerations in experimental stress analysis is the application of load to the element under test. In cases where full-size structures are being tested under service conditions the need, of course, does not arise, but in general the means of applying loads accurately and conveniently in any experiment has to be thought of in relation to the method of analysis and the selection of materials from the earliest stages of the project. Three aspects of the problem can be distinguished: firstly, the means of creating the required forces; secondly, the means of applying these forces to the test piece and, finally, the means of measuring the forces.

METHODS OF GENERATING FORCES

The methods of generating forces may be classified in the following manner:

A. Static or Slowly Varying Forces

(i) Dead load.
(ii) Dead load with levers.
(iii) Screw jacks.
(iv) Hydraulic jacks.
(v) Pneumatic devices.
(vi) Centrifugal force.

B. Dynamic Forces

(i) Mechanical inertia systems.
(ii) Hydraulic jacks.
(iii) Electro-magnets.

These methods cover the majority of loading devices in common use although it is possible to think of others for rather unusual circumstances such as the use of rockets for applying forces to a high tower.

Where it can be used, direct dead loading is simple and accurate but it is, perhaps paradoxically, limited to tests on very large and on very small structures as in these two cases the bulk of the material required to make up the load is less likely to be inconvenient. The use of dead load reduces the complications of applying and measuring loads to a minimum and is particularly indicated in tests in which the load has to be held constant for prolonged periods, for example where creep strains are being measured. Where it is impossible to apply a sufficient amount of load directly to the test piece, a lever system may be employed, but this, of course. increases the complications of the supporting framework and may become impracticable if many loads have to be applied.

Screw jacks are useful in experiments where a controlled rate of straining is required but, except for small rigs, are rather inconvenient. Hydraulic rams are widely used both in testing machines and as separate jacks in test rigs, the ram load being measured by one of a number of devices described below. The fineness of control of hydraulic systems and the possibility of simultaneous operation of hydraulic rams from a central pump station are important advantages. Hydraulic rams can be made in practically any size to apply loads of anything from a few Newtons to hundreds of tonnes and can be used to apply both static and varying loads.

Loading by air pressure is occasionally useful(1) as an alternative to hydraulic loading, but should be restricted to small systems because of the considerable amount of energy stored by compressed air.

Loading by centrifugal force is employed in testing structures and components in high acceleration fields. In practice this is achieved by use of a large centrifuge. In civil engineering, this method has been developed to simulate gravitational loading of dams [(2)] and other structures in which self-weight stresses are important.

It is frequently necessary in studying the dynamic behaviour of structures to apply a cyclic force, the frequency of which can be varied. This is done in order to determine modes of vibration and critical frequencies. As a rule, quite small forces are sufficient for these tests. A well known mechanical device consists of two contra-rotating unbalanced masses as shown in Fig. 2.1; as will be seen, the centrifugal force resulting from the rotation of one of the masses alternately reinforces and opposes the force produced by the other; thus the device applies an oscillating vertical force to the element to which it is attached. The unbalanced masses can be driven by a variable speed electric motor through reduction gearing or by a fixed speed motor and a hydraulic variable speed device.

Electromagnetic vibration exciters are also available. A typical cross-section of such a device is shown in Fig. 2.2. These can be made in sizes similar to a radio loudspeaker movement, capable of applying forces of a few Newtons, up to very large units which develop forces of several hundred Newtons. The variation of the force is controlled by a suitable electronic function generator the signal from which is amplified and applied to the armature windings.

Hydraulic rams can also be employed to supply pulsating loads, but the control is much more complicated. On the other hand, low frequency loading is more easily arranged in this way especially when large forces are required. Thus hydraulic loading is used on machines for fatigue tests on large structural elements.

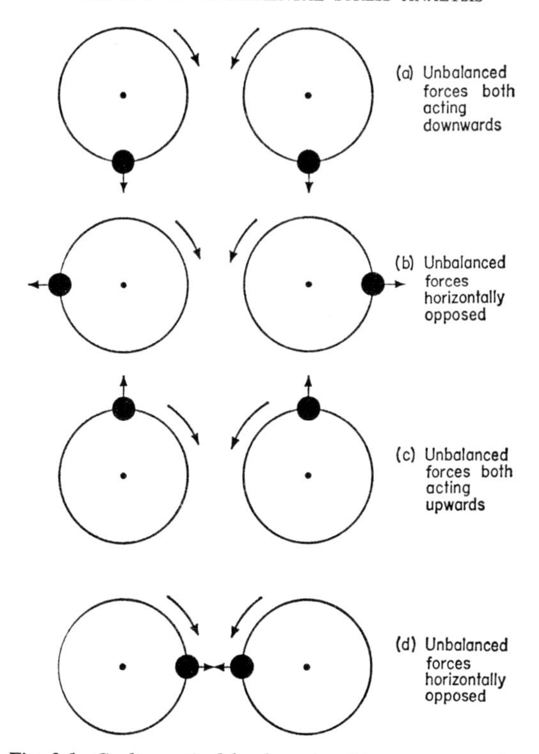

Fig. 2.1. Cyclic vertical load produced by contra-rotating unbalanced masses.

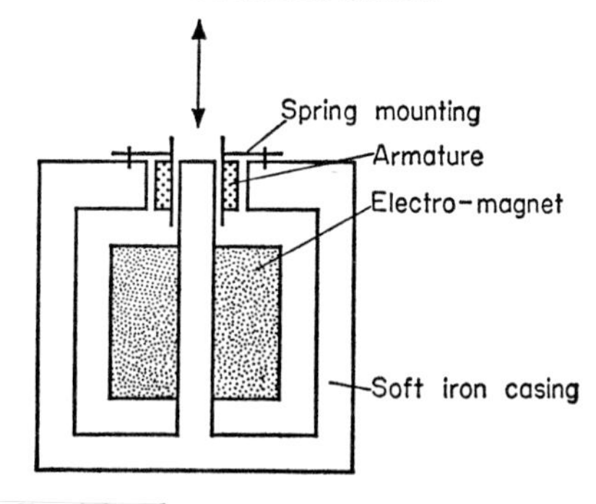

Fig. 2.2. Electromagnetic vibration generator.

Suddenly applied and impact loads are most easily obtained by falling weights or pendulums.

METHODS OF APPLYING FORCES

Essentially the systems for applying load to a test piece fall into two classes:

(*a*) Self straining rigs.

(*b*) Arrangements in which forces are transmitted directly into the ground or the laboratory floor.

The first category includes testing machines and frameworks which merely rest on the floor or bench without transmitting any loads to it other than the weight of the equipment. The second category includes "strong floor" systems and other arrangements in which the "ground" provides the reactions to the applied loads. (Descriptions of typical loading rigs and installations will be found in references 3–6.) Although testing machines are limited in their use to the loading of comparatively simple elements and connections, they are essential equipment in an experimental stress analysis laboratory especially for calibration experiments and tests on materials. A wide range of testing machines is available commercially capable of applying static or pulsating loads of up to 1000 tonnes in tension or compression. Machines are also available for testing beams and torsion members. The majority of modern testing machines employ hydraulic rams for the application of loads. In some it is possible to connect portable rams for applying loads to more complicated structures and in such a case it is, of course, necessary to employ a suitable framework to provide the reaction to the jack load. Certain machines are arranged to impose a controlled deformation on the specimen and to measure the corresponding load.

Special rigs are essential for testing complicated structures which cannot be set up in standard testing machines. The degree of elaboration of a test rig can often be substantially reduced if floor anchorages are available, but these are expensive to provide for large loads and can only be installed as part of a building.

METHODS OF MEASURING FORCES

Fundamentally, the methods available for measuring loads applied to test pieces and structures are rather limited in number. They include:

(*a*) Reference to a mass standard.
(*b*) Elastic deformation devices.
(*c*) Piezoelectric devices.
(*d*) Photo-elastic balances.

If dead load is employed, either directly or through a lever system, the reference to a standard of mass is obvious, and on reflection it will be found that the calibration of devices in categories (*b*), (*c*) and (*d*) will ultimately go back to the same standard. Apart from the direct use of dead load, some commercial testing machines use weigh beams or pendulum dynamometers to measure the load applied by a screw jack or hydraulic ram. The pendulum dynamometer, shown in outline in Fig. 2.3, provides a most accurate means for the measurement of the pressure in a hydraulic system. In testing machines employing this system, the hydraulic rams are specially designed to eliminate friction.

Elastic deformation devices for load measurement are very numerous and very diverse.[7] Under this heading we may include:

(i) Spring balances (helical spring).
(ii) Proving rings (diametral deflection of a steel ring).
(iii) Elastic weighing beams (deflection of a steel beam).
(iv) Load cells or weigh bars (compression or tension of a steel block or bar).

(v) (*a*) Measurement of hydraulic or pneumatic pressure in a ram system by Bourdon gauge.
(*b*) Pressure capsules (measurement of oil pressure in a closed cell by Bourdon gauge).
(vi) Volumetric change load cell.

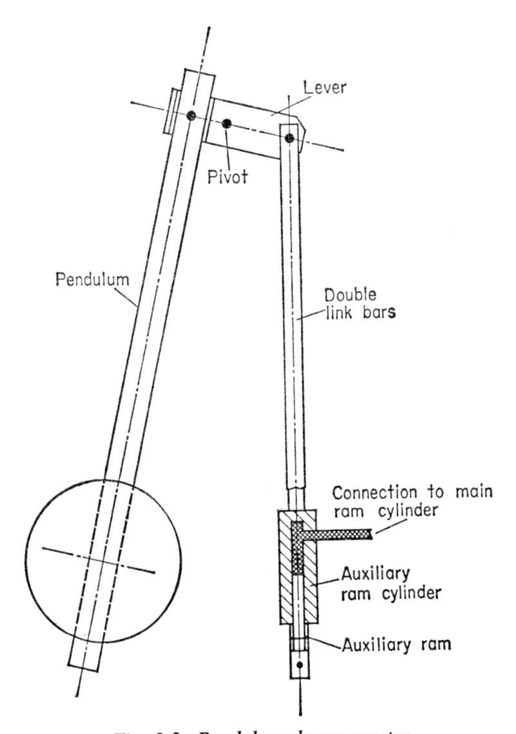

Fig. 2.3. Pendulum dynamometer.

All these devices depend on the measurement of the deformation of an elastic element. In a spring balance the deformation is so large that it can be observed directly on a scale or after simple mechanical magnification. In the devices included under (ii), (iii) and (iv), the deformation is rather small and may be measured with a dial gauge, an electric inductance, capacity or resistance strain gauge transducer as may be convenient. The use of electrical measuring devices means that variable loads can be recorded along with strain or other data and that loads can be measured in inaccessible places. Spring balances are suitable for load measurement from a few Newtons up to about 2 tonnes. Proving rings are practicable between a few Newtons and 200 tonnes, whilst weigh beams can be contrived to deal with practically any load. Load cells and weigh bars usually depend on strain measurement by electrical resistance strain gauges. They are compact and have practically no upper limit for load measurement; the lower limit is set by practical considerations of size and the sensitivity of the strain gauge equipment[8, 9].

Category (v) refers to the use of Bourdon gauges either to measure the oil or air pressure in a loading system or in a closed, flexible cell interposed between the loading device and the test piece. (It will be realised, of course, that a Bourdon gauge is an elastic deformation device.) Pressure gauge systems can be devised to deal with almost any range of total force, but must be used with caution with hydraulic rams where friction may be considerable.

The last type of load cell consists of a vessel full of liquid and fitted with a capillary tube and micrometer screw as shown in Fig. 2.4.

As the cell is loaded, its volume alters and the liquid level in the capillary tube alters; it is restored to its original level by adjusting the micrometer screw, the reading of which is calibrated in terms of load. Temperature alterations must of course be avoided whilst readings are being taken.

Piezo-electric devices make use of the fact that the electrical properties of certain crystals are sensitive to pressure. They have been used particularly for high speed engine indicators, but could of course be employed to measure pressure in any hydraulic system and would be particularly applicable to systems in which the pressure fluctuates rapidly.

A photo-elastic balance consists of a simple member of convenient form in which the stresses can be easily calculated. When this member is interposed between the straining device and the model, the force transmitted can be deduced from the fringes observed in the balance member in polarised light. Such an arrangement is of course only likely to be useful in photo-elasticity.

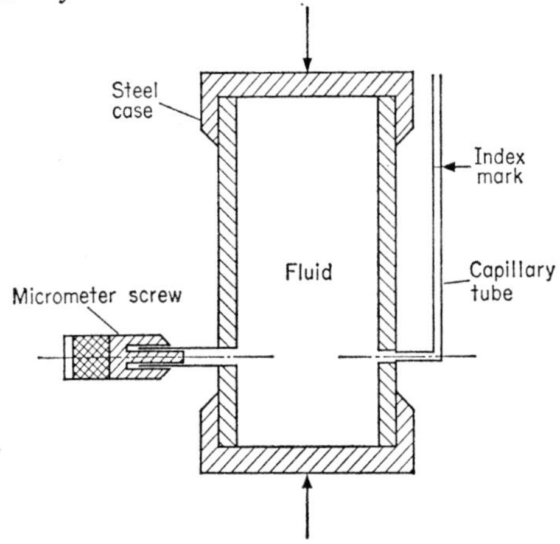

Fig. 2.4. Fluid load cell.

BIBLIOGRAPHY

1. A. J. DURELLI *et al.*, Device for applying uniform loading to boundaries of complicated shape, *Proc. Soc. Exp. Stress Anal.*, **11** (1), 55–64 (1953).
2. A. W. HENDRY, Photoelastic experiments on the stress distribution in a diamond head buttress dam, *Proc. Inst. C.E.*, Part 1, 370–396 (May 1954).
3. L. BAES and Y. VERIVILST, A large mechanised installation for duration tests, *Proc. Soc. Exp. Stress Anal.*, **16** (1), 39–56 (1958).
4. ANON., A 150-ton structures testing machine, *Engineer*, **184** (4796), 594–6 (1947).
5. E. WENK, A frame for testing structural models, *Proc. Soc. Exp. Stress Anal.*, **7** (1), 67–78 (1949).
6. E. L. CLARK, Experimental verification of mathematically derived formulae, *Proc. Soc. Exp. Stress Anal.*, **16** (1), 57–68 (1958).

7. B. Swindells and J. L. Evans, *Measurement of Load by Elastic Devices*, Nat. Physical Lab., Notes on Applied Science, No. 21. H.M.S.O., 1960.
8. G. Fouretier, *La dynamometrie de precision*, Selected papers on stress analysis, Institute of Physics, pp. 92–8. Chapman & Hall, London; Reinhold Pub. Corp., New York, 1959.
9. D. S. Dean, An improved strain gauge type of load cell thrust transducer, *Aero. Res. Coun., London, R. and M.* 3153 (1960).

III

MECHANICAL, OPTICAL AND OTHER GAUGES

ELASTIC strains, even in materials having low elastic moduli, are very small and in order that they may be measured it is necessary to magnify them in some suitable manner. For example, suppose that it is required to construct a strain gauge in which the gauge length is to be 250 mm and the least scale division, of length 1 mm is to be equivalent to the strain produced by 1 N/mm^2 tension or compression in steel, i.e. approximately 500×10^{-6}. To do this a magnification of 1250 : 1 would be required. It will be appreciated that it is not easy to meet this specification by purely mechanical means, whilst at the same time producing an instrument sufficiently robust for practical purposes(1). The problem is of course a good deal easier if the gauge length is longer, but in practice, this is possible only in certain types of work as, for example, in taking measurements on full size structural members. If a complex stress distribution has to be explored it is usually necessary to employ a short gauge length instrument to avoid errors through averaging across a stress gradient.

The deflection or displacement of a structure or element may be greater than the strains from which they result by several orders of magnitude so that their measurement is correspondingly easier. Any necessary magnification can be achieved mechanically or optically, or the movement may be measured directly by vernier or micrometer devices.

In the following pages we shall consider some of the more commonly used instruments for measuring deflection and strains by non-electrical means.

DIAL GAUGES

By far the most convenient device for measuring small displacement of models or full size elements subjected to "static" loading is the dial gauge. In this instrument the movement of a plunger placed in contact with the test piece is transmitted by a rack and pinion to a gear train and pointer. The plunger is spring loaded to keep it against the test piece. According to the magnification produced by the gear train, the scale is graduated in 0·01 or 0·001 mm divisions. The scale can be rotated to set the pointer at zero at the commencement of a test. The calibration of dial gauges is easily checked using a micrometer screw gauge and a suitable jig. Owing to friction at the many moving surfaces, there is a tendency for dial gauges to stick, particularly if the spring return is relied upon to make the gauge follow the movement of the test piece; to prevent this the gauge should be tapped lightly before taking a reading.

Dial gauges are provided with a mounting lug by which they can be fitted to a suitable stand; magnetic stands are particularly useful for setting up dial gauges as they frequently obviate the necessity for special mountings.

Although the pressure required to move the plunger of a dial gauge is quite small, it may not be negligible when dealing with light plastic models. For example, in a typical 25 mm travel dial gauge, the spring constant and frictional force in the mechanism may be of the order of 0·02 N/mm and 0·25 N, respectively. If the force required to operate a dial gauge is excessive, it will be necessary to measure the deflections by telescopes or microscopes which of course impose no loading on the model. Similar devices may be required in measuring the deflections of large structures as it may then be impracticable to set up an independent framework to support dial gauges. The use of dial gauges for the measurement of dynamic deflections is discussed in Chapter X.

TELESCOPES AND MICROSCOPES

Several kinds of telescopic and microscopic devices are useful for measuring displacements. The simplest device is the ordinary surveying level reading against a scale attached to the structure under test. More elaborate instruments include cathetometers, travelling microscopes, and micrometer eyepiece microscopes. The selection of the most appropriate instrument depends on the magnitude of the displacement to be measured and the distance of the instrument from the point at which the displacement is being measured.

Rotational displacements can be measured very accurately using a theodolite, mirror and scale set up as shown in Fig. 3.1. If the reading of the theodolite cross-hair on the scale alters by an amount d which is small compared with the distance of the scale, L, from the specimen, the angle of rotation of the latter is equal to $d/2L$ radians. The line of sight of the instrument rotates by an amount equal to twice the rotation of the specimen. An alternative to a theodolite is to use a lamp with a collimated beam which projects a reference image on a scale.

THE MOMENT INDICATOR

Although deflection measurements are often important, they do not as a rule afford much information about the stress conditions in an element. In rigid frame structures, however, it is possible to deduce bending moments at sections of members by means of a device known as the moment-indicator [(2)].

This is a neat application of the slope-deflection equation

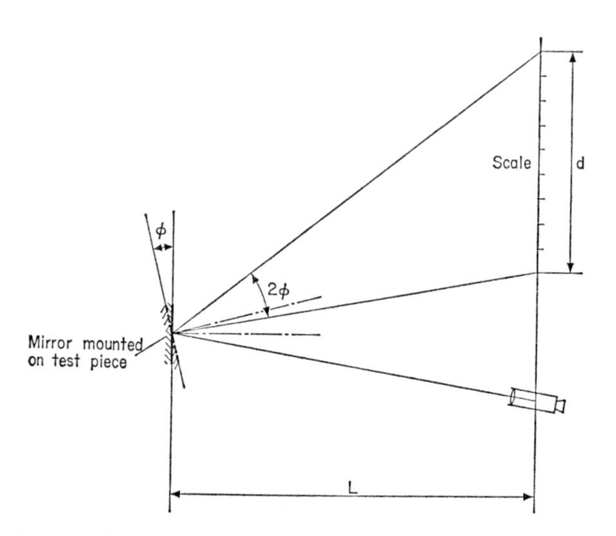

Fig. 3.1. Measurement of angle of rotation. For small angles $\phi = d/2L$ radians.

referring to Fig. 3.2, if A and B are two sections on a beam member, the bending moment at A is given by:

$$M_{AB} = \frac{2EI}{L}\left(2\theta_A + \theta_B - \frac{3\delta}{L}\right) \tag{3.1}$$

If the displacements of points a and b on arms attached to the member at A and B respectively, are as shown in Fig. 3.2, it will be seen that the movement of a is:

$$aa' = \frac{2L}{3} \cdot \theta_A - \delta \tag{3.2}$$

and of b:

$$bb' = \frac{L}{3} \cdot \theta_B \tag{3.3}$$

The relative movement of a and b is thus:

$$aa' + bb' = \left(\frac{2L}{3} \cdot \theta_A - \delta + \frac{L}{3} \cdot \theta_B\right)$$

$$= \frac{L}{3}\left(2\theta_A + \theta_B - \frac{3\delta}{L}\right) \tag{3.4}$$

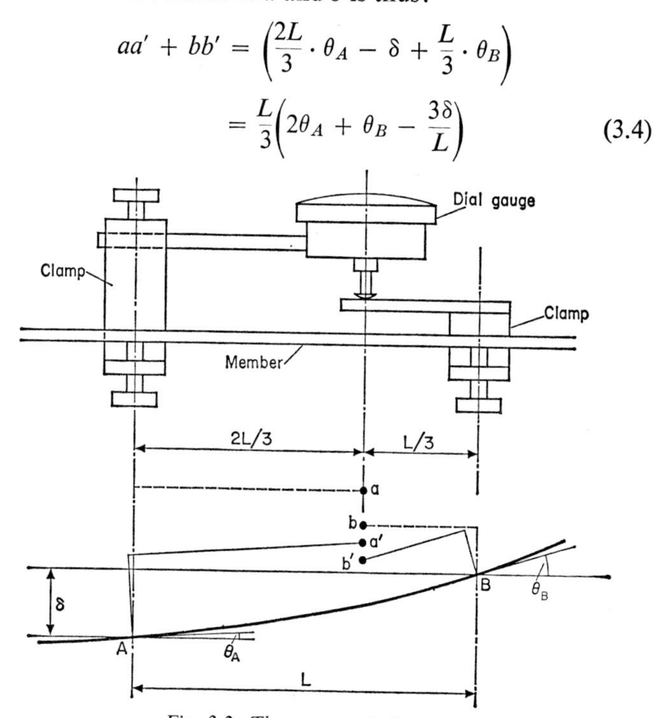

Fig. 3.2. The moment indicator.

Comparing this with the slope deflection equation, it is evident that M_{AB} will be obtained by multiplying the relative displacement by $6EI/L^2$. The sign of the moment can be found by reversing the instrument so that the measurement is taken at $L/3$ from A. In practice the displacement is conveniently measured by means of a dial gauge as indicated in Fig. 3.2. This is a very simple device and is readily adaptable to tests both on models and on full scale structures.

MECHANICAL STRAIN GAUGES

The general problem of mechanical strain gauges has been

outlined at the beginning of this chapter. As suggested there, these instruments fall into two broad classes: (*a*) short gauge length, high magnification strain gauges and (*b*) long gauge length, low magnification instruments. Although there is available a wide variety of strain measuring instruments for use in materials testing, there are very few gauges of the first class which are suitable for experimental stress analysis. Of those available, the best known instrument is perhaps the Huggenberger Tensometer which combines magnifications of up to 2000 with reasonably robust construction. Gauge lengths of 10 mm or 20 mm are available. The principle of the instrument can be seen from Fig. 3.3; as may be seen from this diagram,

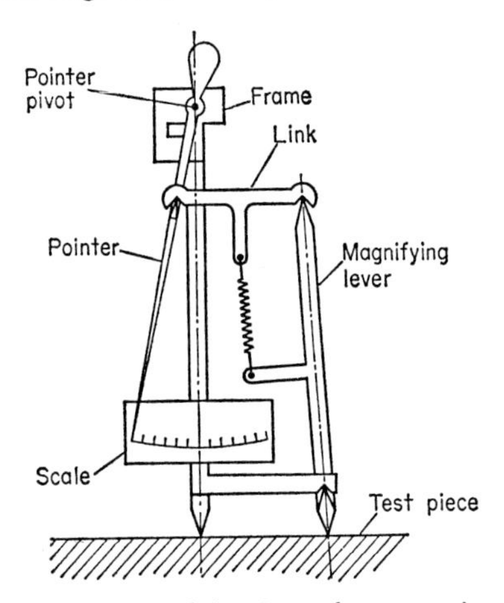

Fig. 3.3. Mechanism of the Huggenberger strain gauge.

the strain is magnified by a double lever system and the reading is indicated by a pointer moving across a scale. The instrument is held against the test piece by spring pressure or a suitable clamp. Excellent results are obtainable with Huggenberger gauges, but some experience in using them is necessary [3].

Another instrument available with short gauge length is the Johanssen gauge in which the magnifying element is a twisted metal strip, fixed at one end and attached to a movable knife edge at the other. A light pointer attached to the twisted strip moves over a scale when the gauge points move relative to one another. A large and a small version of the instrument are produced, the former with a gauge length adjustable between 50 mm and 250 mm and the latter between 3 mm and 11 mm. The gauge is light and robust in construction and very high magnifications are possible [4].

Several types of long gauge length instruments are available for measurements on large structures [5, 6, 7]. These employ a dial gauge with or without a simple lever magnification system. A dial gauge reading to 0·0025 mm has a mechanical magnification of about 500 and it is possible to read strains of, say, 10×10^{-6} on a 250 mm gauge length without any additional amplification. With a lever system, smaller strains can be detected or this strain can be measured on a shorter gauge length. Gauges of this type are thus useful for strain measurements in concrete and masonry specimens and structures and, if the gauge length is long enough, on large steel members.

Several gauges of this type are available. The well known Whittemore gauge has no levers and is intended for measurement on concrete members on a 250 mm gauge length. The gauge shown in Fig. 3.4 has been described by Jenkins[5] and used by him for measurements on steel structures.

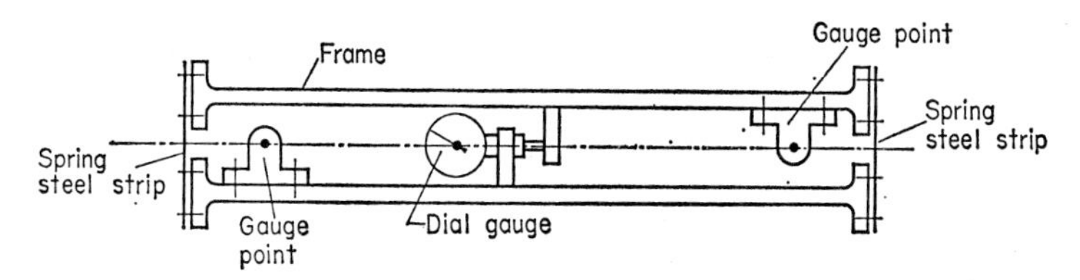

Fig. 3.4. Jenkins' strain gauge.

The "Demec" gauge developed by the Cement and Concrete Association [6], is shown in Fig. 3.5.

It is normally made with a 50 mm, 100 mm or 150 mm gauge length and consists of a Invar main beam with two conical gauge points, one fixed at one end and the other pivoting on a knife edge. The pivoting movement is transmitted to a 0·0025 mm dial gauge suitably mounted on the beam. Provision is made to compensate for thermal movement and only small temperature corrections are necessary in use. The gauge points are located by small, drilled, stainless steel pads cemented to the structure; the correct centre distance is ensured by use of a

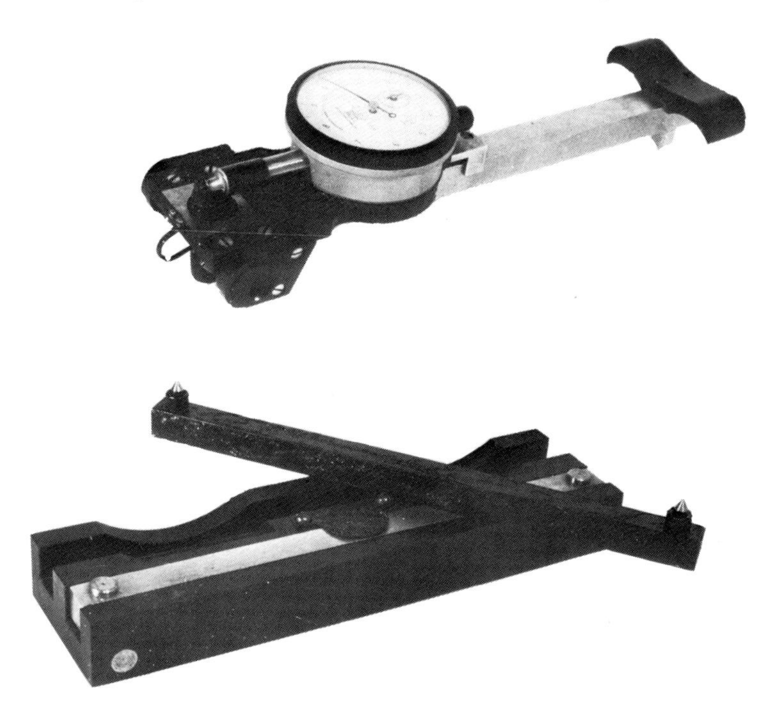

Fig. 3.5. Demec gauge. The lower photograph shows the calibration and setting out bars.

suitable setting out bar. The makers state that readings can be obtained with an accuracy of $\pm 3 \times 10^{-6}$. The instrument is particularly suitable for observations on concrete structures and elements.

THE DE LEIRIS PNEUMATIC STRAIN GAUGE

An interesting and effective method of obviating the difficulties of mechanical magnification is used in the de Leiris strain gauge [8, 9]. The principle is shown in Fig. 3.6: as will be seen from this diagram, air from a constant pressure vessel is discharged from an orifice (*A*), the flow from which is controlled by a small plate (*B*). The orifice and the plate are attached to the test specimen at the gauge points (*P*) and (*Q*) respectively, and the flow through the orifice is measured by a manometer (*C*) connected across a second orifice (*D*). If the strain in the specimen is tensile, plate (*B*) moves away from the orifice (*A*) and the flow increases and vice versa if the strain is compressive. Change in the flow results in an alteration in the loss of head across the orifice (*D*). The air supply pressure is maintained at a constant head (*H*) by means of a dip tube bubble tank (*E*); this pressure is usually about 1 m of water. A number of gauge points can be operated from a common air supply, each gauge requiring, of course, its own control orifice and manometer tube.

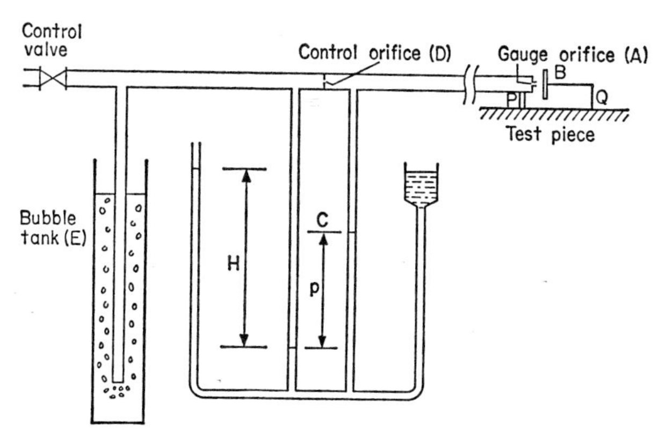

Fig. 3.6. Principle of pneumatic strain gauge.

Assuming incompressible flow and considering the flow through the two orifices, it is easily seen that:

$$C \,.\, A_c \,.\, \sqrt{(2g \,.\, p)} = C \,.\, A_m \,.\, \sqrt{[2g \,.\, (H - p)]} \qquad (3.5)$$

where:

C is the coefficient of contraction, assumed to be the same for each orifice;
A_c is the area of control orifice;
A_m is the area of measuring orifice;
g is the gravitational acceleration;
H is the working pressure;
p is the pressure drop across the control orifice;

thus:

$$p = \frac{H}{1 + (A_c/A_m)^2} \qquad (3.6)$$

In the above, H and p are measured in terms of heights of the liquid columns in the gauge tubes.

The area of the measuring orifice is $A_m = \pi \,.\, d \,.\, s$, where s is the clearance between the diaphragm and the plate (B), s being small compared with d. It is clear from this that over the whole possible flow range, i.e. from the measuring orifice being closed to fully open, the ratio of the pressure drop across the control orifice to the working pressure, i.e. (P/H), is not directly proportional to the change in the orifice area and thus to the relative movement of the gauge points, but if the ratio A_m/A_c is about 0·6, the relationship is very nearly linear for the small range of movement associated with elastic strains.

The pneumatic gauge is capable of very large magnifications under favourable conditions (as great as 100,000:1) and has a number of advantages, not the least of which is the possibility of taking readings at a control instrument from a number of gauge points located at a distance from it. The gauging heads and measuring instruments can be robustly constructed at reasonable cost and can be adapted to measure displacements and pressures as well as strains.

The development of this gauge slightly preceded that of

electrical strain gauges, and it seems to have been overshadowed by the latter; in recent years, however, it has been developed for precision gauging in metrology [10, 11] and has been used for pressure cells in soil mechanics [12]. Its advantages are such that it merits serious consideration for experimental stress analysis.

BIBLIOGRAPHY

1. A. F. C. POLLARD, The mechanical amplification of small displacements, *J. Sci. Instrum.*, **15**, 37–55 (1938).
2. A. C. RUGE and E. O. SCHMIDT, Mechanical structural analysis by the moment indicator, *Proc. Am. Soc. Civil Eng.* (1938).
3. T. R. CUYKENDALL and G. WINTER, Characteristics of the Huggenberger strain gage, *Civ. Eng.* (*New York*), **10** (7), (1940).
4. J. M. HAWKES and H. FEALDMAN, The measurement of stresses in framed structures, *Civ. Eng. and Pub. Wks. Rev.*, **47** (1952); **48**, 75–7, 166–8, 261–3 (1953).
5. R. A. SEFTON JENKINS, *Testing of Prestressed Steelwork*, Prelim. vol., Conf. on the Correlation between Calculated and Observed Stresses in Structures, pp. 44–53. Inst. C.E., London, 1955.
6. H. L. WHITTEMORE, The Whittemore strain gauge, *Instruments*, **1** (6), 299–300 (1928).
7. P. B. MORICE and G. D. BOSE, The design and use of a demountable mechanical strain gauge for concrete, *Mag. Conc. Res.*, **5** (13), 37–42 (1953).
8. H. DE LEIRIS, A new pneumatic amplification method and its application to extensometry, Proc. 7th Int. Cong. App. Mech., 4, pp. 121–127, 1948.
9. H. DE LEIRIS, Sur la mesure des constantes elastiques par amplification pneumatique des deformation, Proc. 5th Int. Cong. App. Mech., pp. 193–197, 1938.
10. J. C. EVANS, The pneumatic gauging technique and its application to dimensional measurement, *J. Inst. Prod. Eng.*, **36**, 110.
11. J. C. EVANS, Pneumatic gauging techniques, *Research*, **11**, 90–97 (1958).
12. P. WIKNAMARATNA, A new earth pressure cell, Proc. 5th Int. Conf. for Soil Mech., pp. 509–512.
13. K. F. SMITH, Types of strain measuring devices and their range of utility, *Product Engineering*, **18** (1) (1947).
14. K. SWAINGER and J. TWYMAN, An optical rectangular rosette extensometer for large strains, *J. Sci. Instrum.*, **25** (6), 187–9 (1948).

IV

THE ELECTRICAL RESISTANCE STRAIN GAUGE

THE principle of the electrical resistance strain gauge was discovered by Lord Kelvin when he observed that the resistance of a wire alters when it is subjected to stress. However, no use was made of this principle for strain measurement until the 1930's when it was rapidly developed into one of the most useful tools in experimental stress analysis.

BASIC PRINCIPLES

A typical strain gauge element consists of a continuous length of resistance wire of about 0·025 mm diameter wound as indicated in Fig. 4.1 and cemented to a paper backing.

Gauges are also available in which the conductor takes the form of a metallic foil "ribbon", 0·01 mm to 0·025 mm thick,

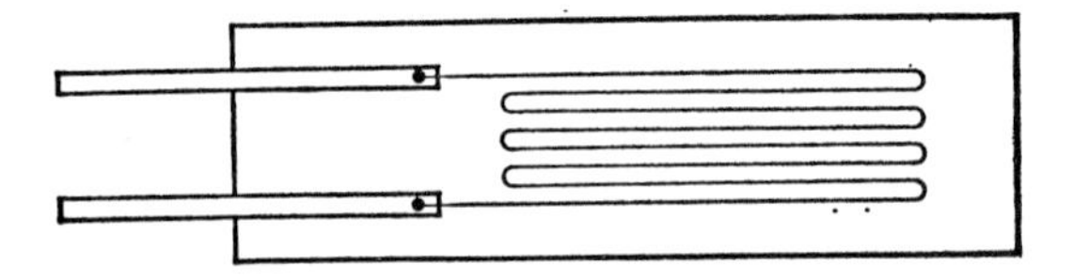

Fig. 4.1. Wire-wound strain gauge element.

mounted on a thin epoxyethylene backing. The strain gauge is cemented to the material in which it is required to measure the strain. As the test specimen extends or contracts under stress in the direction of the windings, the length and cross-sectional area of the conductor alter, resulting in a corresponding increase or decrease in electrical resistance. The alteration in resistance is, within wide limits, proportional to the strain in the specimen, that is:

$$\epsilon = \frac{1}{k} \cdot \frac{\delta R_0}{R_0}$$

where k is termed the gauge factor and R_0 is the gauge resistance. On the basis of change of length and cross-sectional area only, it can be shown that the gauge factor would be $(1 + 2\nu)$, where ν is Poisson's ratio. However, for most materials the specific resistance of the conductor changes with strain and the gauge factor is usually about 2·0.

It will be appreciated that although a gauge of the type shown in Fig. 4.1 is intended to measure strains in the direction of its length only, the small lengths of the conductor in the direction at right angles to the axis of the gauge impart some degree of sensitivity to transverse strain; fortunately, this effect is small enough to be neglected in work of normal accuracy, amounting to 2 per cent of longitudinal sensitivity in wire wound gauges and about one tenth of this in foil gauges. It will also be noted that a linear gauge measures normal strains only: shear strains must be deduced from gauge readings by the methods described in Chapter VI.

Both wire-wound and foil strain gauge elements are available commercially covering a very wide range of applications. Single gauges are made in a variety of resistances, dimensions and materials as are rosette gauges having windings in two or three directions to each other, as typified in Fig. 4.2.

As regards wire wound gauges, those most frequently employed are wound with nichrome or Eureka wire and bonded with cellulose cement to a paper backing. For most purposes a resistance of 120 or 200 ohms is suitable. These gauges have a gauge factor of about 2·0. A higher gauge factor (of about 3·0) is obtainable with stainless steel wound gauges; these gauges also show lower hysteresis, but are much more expensive than the nichrome type. Other gauge wire alloys

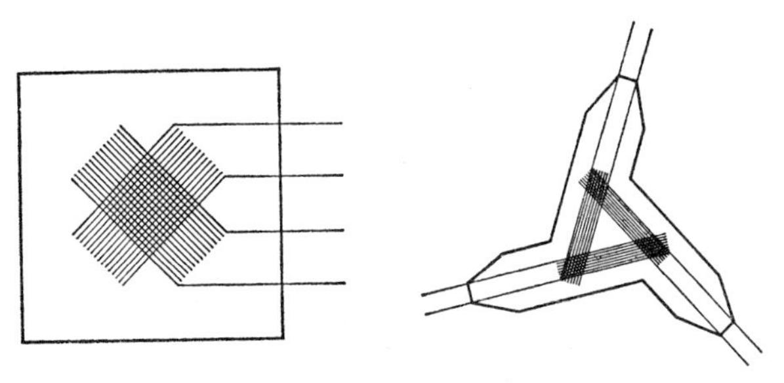

Fig. 4.2. Typical rosette strain gauges.

such as Karma have lower temperature coefficients than nichrome. Swainger [5] has reported that a copper–nickel alloy, "Minalpha" is suitable for measuring large strains.

Gauges bonded with bakelite or epoxy resin instead of cellulose cement are available for strain measurements at temperatures up to about 200°C, or in cases where stability over a long period of time is essential.

Foil gauges are made in the same basic grid patterns as wire wound gauges and, in addition, in a variety of special types which would be difficult to produce in wire wound form. The ribbon element of a foil gauge, which may be of cupro-nickel or ni-chrome alloy, has certain advantages over a wire element in that it permits better heat dissipation from the gauge and gives better bonding to the backing. This improved bonding results in a higher gauge factor and also permits higher shear force to be transmitted. Foil gauges have high current carrying capacity and the sensitivity of a foil gauge to strain at right angles to the principal axis is much less than in the case of a wire wound gauge. Greater stability is also claimed.

Amongst the special types of foil gauges which have been developed are diaphragm gauges for use in pressure measuring devices, torque gauges and long linear gauges for measurements on concrete. Gauges are available for use at temperatures up to 800°C and some have very low temperature coefficients so that they can be used without a compensating gauge.

STATIC STRAIN MEASUREMENT

The usual method of measuring the change of resistance in a gauge element is by means of a Wheatstone bridge circuit, as shown in Fig. 4.3. In this diagram the bridge arms R consist of equal resistances similarly constructed so that any thermal changes will be the same in both. The two other arms of the bridge are strain gauge elements. Normally, one of these is mounted on the test specimen and the other is mounted on a piece of unstrained material. The object of the second gauge, termed a "dummy gauge", is to eliminate any unbalance of the bridge arising either from self heating or changes in ambient temperature. It is therefore necessary that the dummy gauge should be subject to the same conditions as the active gauge, ideally mounted on an identical, but unstrained, specimen, kept close to actual test piece. The temperature coefficient of resistance of strain gauge elements is typically of the order of $10\text{–}15 \times 10^{-6}$ per °C which is large enough to make the use of a dummy gauge essential. Special gauges can be obtained having very small temperature coefficients—say 2×10^{-7} per °C and these can be used without temperature compensation under certain circumstances. It must also be noted that thermal strains can be set up in a gauge wire if it is mounted on material of different coefficient of thermal expansion; it is thus essential for dummy gauges to be mounted on the same kind of material as the measuring gauges as well as being kept under the same ambient conditions.

As it is impossible to manufacture strain gauge elements to

have exactly the same resistance to within, say, 0·001 ohm which would be necessary to obtain initial balance of the bridge, a potentiometer is inserted between the active and dummy gauges. This is known as an "apex resistance" and the galvanometer is tapped into this resistance at the balance point. The circuit in Fig. 4.3 would indicate change of resistance in the active gauge, and thus strain, by galvanometer deflection.

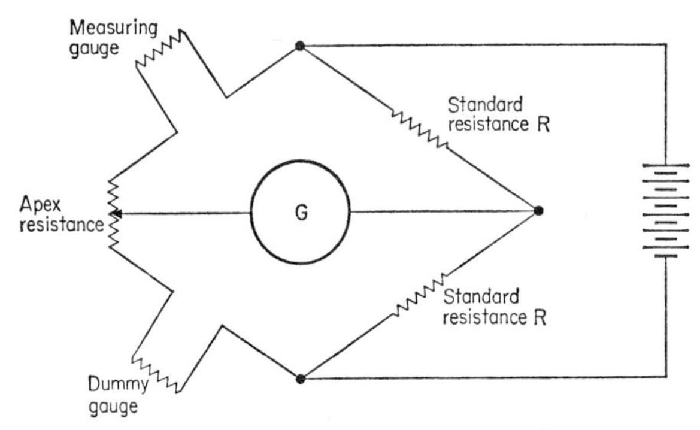

Fig. 4.3. Simple strain gauge circuit.

The galvanometer scale could be calibrated to read strain directly, but in practice provision is necessary for accommodating gauges having different gauge factors; this entails a slight modification of the basic circuit which will be described later.

Another possibility is to place a calibrated slide wire at the apex of the bridge between the standard resistances, as indicated in Fig. 4.4; measurements are then obtained by rebalancing the bridge, as in ordinary Wheatstone bridge practice. If the slide wire is of sufficient length and resistance, it is possible to dispense with apex resistances provided that a zero reading is taken for each measuring gauge circuit.

Stress analysis usually requires strain readings to be taken at a large number of points on the test piece: this is very easily achieved simply by adding a strain gauge–apex resistance–dummy gauge circuit for each point to be gauged. The moving arms of the apex resistances are then connected to a selector switch, as in Fig. 4.4, so that the galvanometer can be connected to each gauge circuit in turn and the reading taken from it as previously described.

Instruments embodying the principles described above using ordinary moving coil galvanometers and slide wires are, of course, suitable only for the measurement of static strains.

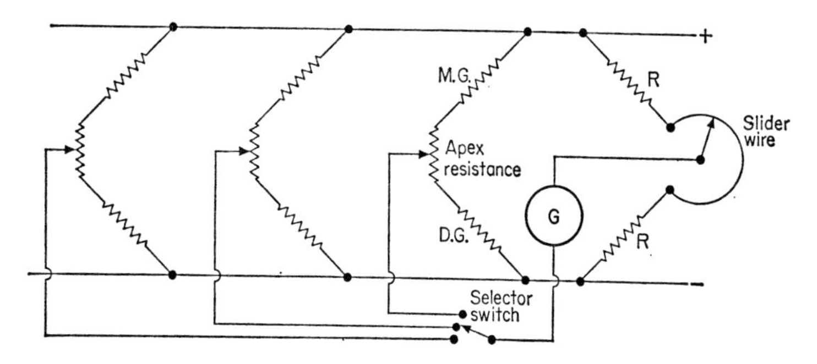

Fig. 4.4. Multiple strain gauge bridge circuit.

Commercially available equipment of this kind usually includes certain elaborations in the circuit; for example, to provide several strain ranges it will be possible to switch one of a number of shunts across the slide wire in a null reading instrument. Direct reading instruments require a standardising circuit to ensure that the supply voltage is correct. It is also possible to incorporate in this an adjustment for gauge factor so that the galvanometer scale can be divided in strain units rather than in percentage change of resistance. A typical circuit, used by Messrs. H. Tinsley in their 4907K instrument, is shown in Fig. 4.5. Here standardisation is effected by applying an unbalance of 0·25 per cent on one of the fixed arms of the bridge; the supply voltage is then adjusted until the galvanometer deflection reads the strain corresponding to this unbalanced resistance. For convenience a range of gauge factors is marked on the strain scale. Clearly, this adjustment must be carried out after the bridge has been balanced by means of the apex resistance and, furthermore, it is essential to ensure that the scale zero coincides with the true null position of the galvanometer. This is checked by reversing the battery polarity when balancing the bridge and also when

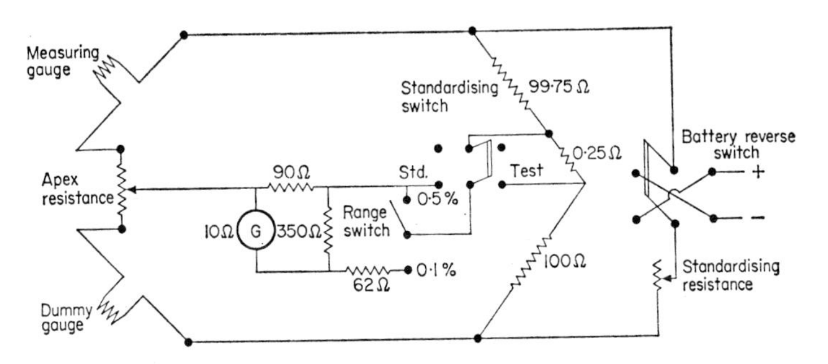

Fig. 4.5. Direct reading strain gauge bridge showing calibration and range circuits (Tinsley).

standardising. In the first adjustment, the galvanometer spot should not deflect when the current is reversed and in the second case the deflection should be the same on each side of the zero. Two range scales are provided by means of alternative shunts across the galvanometer. The least scale reading on the Tinsley instrument is 0·002 per cent (i.e. 20×10^{-6}) on the 0·1 per cent range and the accuracy is stated to be ± 2 per cent of full scale reading.

Greater accuracy is possible with slide wire bridges, although these are slightly slower to operate; instruments of this type manufactured, for example, by Messrs. H. Tinsley and by Messrs. Savage and Parsons are capable of measuring strains to an accuracy of a few micro-inches per inch.

In recent years electronic instruments have been produced for measuring static strains. The most obvious development would be to replace the galvanometer in a simple d.c. bridge by an electronic indicator consisting of a d.c. amplifier and moving coil voltmeter. As compared with a galvanometer, this would have the advantages of greater robustness and would respond more quickly when connected to the bridge; suitable instruments are in fact available for this purpose. d.c. amplifiers are, however, rather more complex than a.c. ones and thus if electronic circuits are introduced it is common practice to supply the bridge with a.c. The use of electronic equipment has the additional advantage that the instrument is then inherently suitable for dynamic strain measurement although additional complications are introduced as capacity unbalance has to be taken into account. An electronic instrument for static work is manufactured by Messrs. Philips which supplies a.c. at 1000 Hz to the strain gauge circuit. The out of balance voltage from the strain gauge bridge is amplified and indicated on a moving coil meter; the bridge is then re-balanced by applying a voltage from a second bridge circuit which is manually adjusted until the meter reading is returned to zero. The adjustment required to the second bridge provides a measure of the strain.

DYNAMIC STRAIN GAUGE EQUIPMENT

The simplest method of measuring rapidly varying strains over short periods of time is shown in Fig. 4.6. As the observations are confined to a short time, temperature compensation

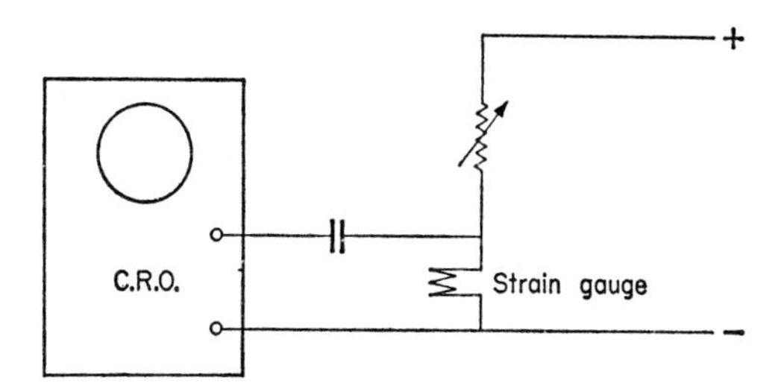

Fig. 4.6. Simple dynamic strain measuring circuit.

is not necessary and it is sufficient to measure the variation of the voltage drop across the strain gauge which is of course directly proportional to the change of resistance. The signal from the gauge can be observed or recorded on a cathode ray oscilloscope, the steady component of the voltage drop being blocked off by a condenser.

A simple d.c. Wheatstone bridge circuit may also be used for dynamic strain measurement, the galvanometer being replaced by a d.c. amplifier and a recorder or oscilloscope. It is essential that the frequency response of the amplifier should be wide enough to deal with the gauge signal. A number of commercially produced oscilloscopes equipped with suitable amplifiers are available; recording can be effected by a suitable oscilloscope camera attachment. Alternatively, a high speed recording galvanometer may be used which may be able to handle signals from as many as twelve gauge circuits simultaneously.

As d.c. amplifiers are comparatively expensive it is common practice to supply strain gauge circuits with alternating current when ordinary a.c. amplifiers can be employed. The circuit shown in Fig. 4.7 can be used, capacity balance being provided by a differential condenser in the portion shown. The output from the circuit will then be a modulated carrier wave which can be recorded on an oscillograph, recording galvanometer or magnetic tape recorder. The carrier wave can be eliminated by a suitable demodulating circuit.

Commercially produced instruments are available such as Philips PR 9300 Direct Measuring Bridge, the arrangement of which is shown in block form in Fig. 4.8.

Two strain gauge elements in a half bridge circuit are supplied with a.c. at 4000 Hz and provision is made for initial balance of the bridge. The out of balance signal is amplified by highly stable amplifiers, the output from which is passed through a band pass filter to eliminate unwanted frequencies, e.g. "noise" from the supply mains. The amplified, modulated carrier wave may be examined on an oscilloscope at output *Bu*,

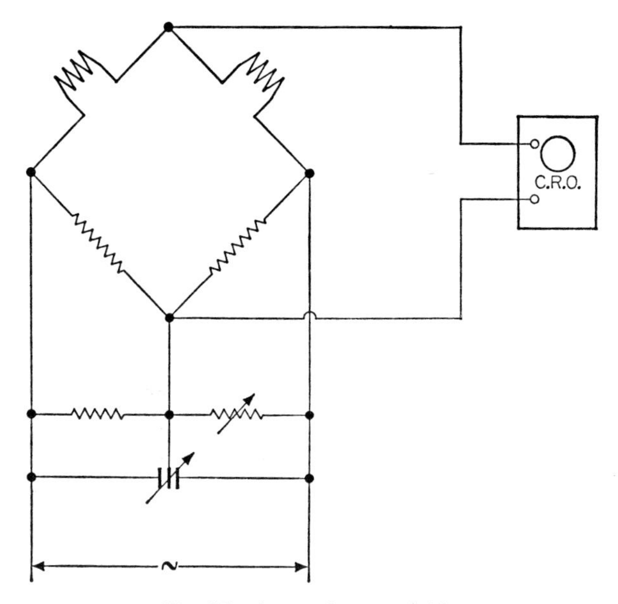

Fig. 4.7. A.c. strain gauge bridge.

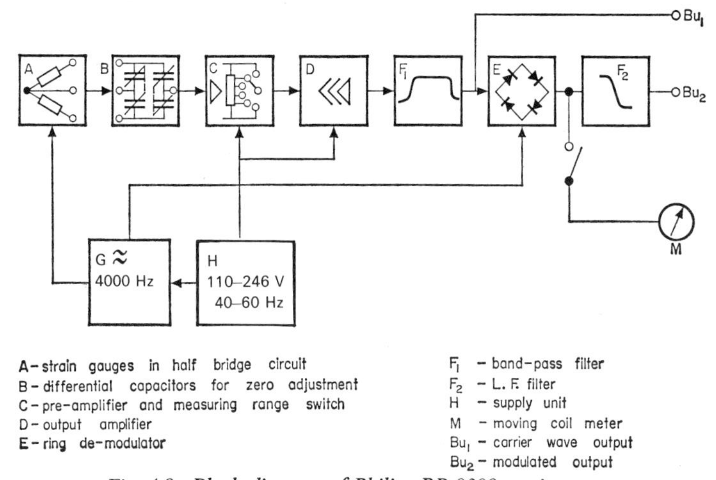

Fig. 4.8. Block diagram of Philips PR 9300 strain gauge.

or the carrier wave may be removed by the ring-demodulator *E* and the gauge signal only is available. The device *F* is a phase discriminator which permits determination of the sign of the signal at Bu_2. Static and slowly variable signals may be measured on a moving coil instrument (*M*). This particular instrument can measure positive or negative strains up to 0·1 per cent on two ranges. On the most sensitive range the least scale division corresponds to 10×10^{-6}. Recording equipment is available for use with this bridge and similar instruments are offered by various other manufacturers.

It should be noted that in a.c. strain gauge work all leads must be fully screened to prevent their picking up external "noise".

MOUNTING AND PROTECTION OF STRAIN GAUGES

A most important procedure in resistance strain gauge work is the mounting of the gauge element on the surface of the test piece. The surface must first be smoothed to grade 0 emery and thoroughly degreased. The procedure varies in detail according to the gauge and cement but, in general, a thin layer of cement is first applied to the surface and allowed to dry out. A second layer of cement is then applied and the gauge pressed into position with a rolling movement so that surplus adhesive is expelled at the ends of the gauge; this is to ensure that the ends of the gauge are firmly secured as otherwise spurious readings will be obtained. Drying out of ordinary cellulose cement under normal conditions may require up to ten days, but this may be reduced to about three days by storage in warm ambient conditions. Very careful local heating with a small electric iron may be employed provided that the gauge temperature is not allowed to exceed 70°C. Rapid hardening cellulose cements are available which will be completely dry in 24 hr; use of these cements will obviously save time, but they are more difficult to use as they begin to dry almost as soon as they are coated on the surface.

Wire wound gauges can be obtained in "self-adhesive" form which means that they are pre-coated with cement and only require brief immersion in acetone before pressing on the prepared surface of the test piece.

Other types of cement are recommended for particular gauges and applications. For epoxy resin bonded gauges and for foil gauges, various Araldite strain gauge cements are recommended. As these cements are unaffected by moisture, their use obviates the necessity for further waterproofing. On the other hand after cellulose cement bonded gauges have been thoroughly dried out, it is essential to protect them from moisture, if measurements are to be taken over a prolonged period. Absorption of moisture by the gauge matrix may result in an apparent reduction of the gauge factor and in extreme cases passage of current through a damp gauge can result in failure of the wire due to corrosion. Ingress of moisture can be prevented by coating the gauge with Vaseline or Di Jell 171. A coating of Araldite is also a possible means of moisture protection and indeed gauges so protected can withstand complete immersion in water for several months.

After mounting and wiring up, each strain gauge element should be checked for resistance before connecting to sensitive equipment. If mounted on a metal test piece, the resistance of each gauge to "earth" should also be checked as the effect of earth leakage can be the same as that of putting a shunt across the gauge. The insulation resistance should be at least 1000 megohms.

Strain gauge elements must be protected from draughts, if necessary by felt pads.

THERMAL DRIFT

The initial balance of a strain gauge bridge will be affected by differential heating of the bridge arms. As previously explained, the use of a dummy gauge opposite the measuring gauge in the bridge is intended to prevent thermal effects due to variations in ambient conditions. Thermal effects also arise due to self-heating of the gauge elements. The latter effect is proportional to the cube of the gauge current; it is thus advantageous to keep this to as low a figure as is consistent with operation of the measuring instruments. For long term measurement 10 mA would be a suitable current for wire wound gauges, but for shorter experiments 25 mA would be permissible for most gauges. Dynamic measurements are unaffected by temperature effects and thus higher currents, say up to 100 mA, may be used for brief periods in short term tests. Very much higher operating currents are permissible with foil gauges: an energy dissipation of 0·015 watt/mm^2 of grid area is suggested by Saunders Roe, implying maximum currents of between 100 and 650 mA.

CALIBRATION

Calibration of strain gauge elements is most conveniently effected by attaching the gauges to a calibration beam strained in pure bending; the surface strain can then be calculated from the dimensions of the beam and the measured deflection. Simultaneous measurement of change of resistance permits determination of the gauge factor. Naturally, this can only be done on a representative sample of the gauges sufficient to establish the gauge factor within reasonable statistical limits. This is normally done by the manufacturer.

Calibration of the strain gauge equipment, in the sense of checking the response of the indicating or recording instrument to a given input signal is usually arranged for by applying a known unbalance to one of the standard arms of the bridge, as in the direct reading instrument shown in Fig. 4.5. Obviously this must be done after the bridge circuit has been balanced.

INSTRUMENTS USING STRAIN GAUGE ELEMENTS

In addition to the measurement of surface strains electrical resistance strain gauges have important applications in devices for measuring such quantities as load, pressure and deflection. A load cell, which might be used for example for measuring the load applied to a test structure or for weighing vehicles, consists of a short pillar of high tensile steel to which strain gauge elements are attached as in Fig. 4.9(a). By connecting

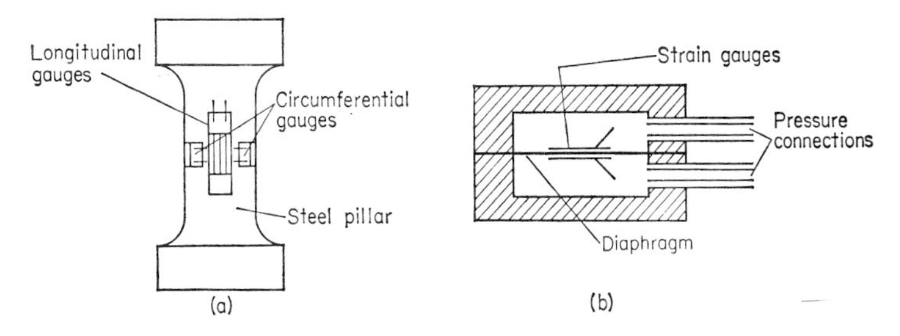

Fig. 4.9. (a) Load cell. (b) Differential pressure gauge.

axial and circumferential gauges on alternate arms of a Wheatstone bridge circuit, bending effects are eliminated and temperature compensation is provided. As long as the stress in the pillar remains within the elastic range a linear relationship will be obtained between load and output from the strain gauge bridge circuit. Load cells of this type can be constructed to cover a very wide range of loads; they are compact and relatively inexpensive. Fluid pressure is conveniently measured by applying the pressure to a diaphragm on one side or both sides of which strain gauge elements have been cemented. A differential gauge can be constructed as indicated in Fig. 4.9(b). In both applications shown in Fig. 4.9, more than one arm of the bridge can be made active. Special care is necessary in mounting the gauges so as to eliminate instability due to moisture. The gauge should be thoroughly dried out in a dessicator before being cemented in position; if cellulose base cement is used, the device should be placed in an oven, the temperature of which is gradually raised to 70°C, and maintained at this level for 24 hr. If hysteresis is observed on loading and unloading during calibration, this is an indication of incorrect fixing of the gauges.

BIBLIOGRAPHY

1. W. B. DOBIE and P. C. ISAAC, *Electrical Resistance Strain Gauges* English Univ. Press, London, 1948.
2. J. J. KOCH *et al.*, *Strain Gauges: Theory and Application.* Cleaver Hume, London.
3. S. F. DOREY, The use of wire-wound resistance strain gauges, *Proc. Inst. Nav. Arch.*, **86**, 61 (1944). [See also *Engineering*, **157** (4089) (1944)].
4. S. C. REDSHAW, The electrical measurement of strain, *J.R. Aero. Soc.*, **428** (50), 568–602 (1946).
5. K. SWAINGER, Electrical resistance wire strain gauges to measure large strains, *Nature*, **159**, 61 (Jan. 1947).
6. C. H. GIBBONS, The use of the resistance wire strain gauge in stress determination, *Proc. Soc. Exp. Stress Anal.*, **1** (1), 41–45 (1943).
7. H. J. GROVER, The use of electric strain gauges to measure repeated stresses, *Proc. Soc. Exp. Stress Anal.*, **1** (1), 110–115 (1943).
8. H. C. ROBERTS, Electric gauging methods: Their selection and application, *Proc. Soc. Exp. Stress Anal.*, **2** (2), 95–105 (1944).
9. L. M BALL, Strain gauge technique, *Proc. Soc. Exp. Stress Anal.*, **3** (1), 1–22 (1945).
10. G. P. TSCHEBATORIOFF, Use of electric resistivity gauges over long periods of time, *Proc. Soc. Exp. Stress Anal.*, **3** (2), 47–52 (1945).
11. E. L. KIMBLE, A method of effecting SR-4 strain gauge operation under water, *Proc. Soc. Exp. Stress Anal.*, **3** (2), 53–54 (1945).
12. W. V. BESSET *et al.*, Improved techniques and devices for stress analysis with resistance wire gauges, *Proc. Soc. Exp. Stress Anal.*, **3** (2), 76–88 (1945).
13. A. G. H. DIETZ and W. H. CAMPBELL, Bonded wire strain gauge techniques for polymethyl methacrylatic plastics, *Proc. Soc. Exp. Stress Anal.*, **5** (1), 59–62 (1947).
14. G. L. ROGERS, Sensitivity chart for wire resistance strain gauges, *Proc. Soc. Exp. Stress Anal.*, **6** (2), 61–63 (1948).
15. E. W. KAMMER and T. E. PARDUE, Electric resistance changes of fine wires during elastic and plastic strains, *Proc. Soc. Exp. Stress Anal.*, **7** (1), 7–20 (1949).

16. E. E. Day and A. H. Sevand, Characteristics of electric strain gauges at low temperatures, *Proc. Soc. Exp. Stress Anal.*, **8** (1), 143–153 (1950).
17. E. E. Day, Characteristics of electric strain gauges at high temperatures, *Proc. Soc. Exp. Stress Anal.*, **9** (1), 141–150 (1951).
18. R. E. Garton, Development and use of high temperature strain gauges, *Proc. Soc. Exp. Stress Anal.*, **9** (1), 163–176 (1951).
19. J. E. Carpenter and L. D. Morris, A wire resistance strain gauge for the measurement of static strains at temperatures up to 1600°F, *Proc. Soc. Exp. Stress Anal.*, **9** (1), 191–200 (1952).
20. H. O. Meyer, A method of waterproofing electrical strain gauges, A letter to the Editor, *Proc. Soc. Exp. Stress Anal.*, **10** (1), 243–245 (1952).
21. F. W. Wells, A rapid method of waterproofing banded wire strain gauges, *Proc. Soc. Exp. Stress Anal.*, **15** (2), 107–110 (1958).
22. R. S. Barker and J. B. Murtland, Protection of underwater SR-4 strain gauge installations on a tunnel liner of a hydro-electric installation, *Proc. Soc. Exp. Stress Anal.*, **14** (2), 131–138 (1957).
23. W. R. Clough *et al.*, The behaviour of SR-4 wire resistance strain gauges on certain materials in the presence of hydrostatic pressure, *Proc. Soc. Exp. Stress Anal.*, **10** (2), 167–176 (1952).
24. G. F. Brosius and D. Hartley, Evolution of high temperature strain gauges, *Proc. Soc. Exp. Stress Anal.*, **17** (1), 67–84 (1959).
25. M. McWhirter and B. W. Duggin, Minimizing creep of paper base of SR-4 strain gauges, *Proc. Soc. Exp. Stress Anal.*, **14** (2), 149–154 (1957).
26. R. I. Brosic and R. C. Dove, Use of electric resistance strain elements in three dimensional stress analysis, *Proc. Soc. Exp. Stress Anal.*, **18** (1), 156–171 (1960).
27. R. C. Dove *et al.*, Selection of gauges for strain measurement at interior points, *Proc. Soc. Exp. Stress Anal.*, **19** (1), 189–190 (1961).
28. C. M. Hathaway, Electrical Instruments for strain analysis, *Proc. Soc. Exp. Stress Anal.*, **1** (1), 83–93 (1943).
29. S. Aughtie, Electrical resistance wire strain gauges—Possible Errors, *Trans. Inst. Mar. Eng.*, **58** (2), 59–66 (1946).
30. H. Majors, Characteristics of wire gauges under various conditions, *Proc. Soc. Exp. Stress Anal.*, **9** (1), 123–139 (1952).
31. R. D. Binns and H. S. Mygind, The use of electrical resistance strain gauges and the effect of aggregate size on gauge length in connection with the testing of concrete, *Mag. Conc. Res.*, **1**, 35–9 (1949).
32. W. T. Walker, Laboratory measurements of stress distribution in reinforcing steel, *Am. Conc. Jour.*, **19** (10), 1041–54 (1948).
33. A. R. Anderson, How to use strain gauges on concrete, *Eng. News Rec.*, **146** (10), 46–49 (1951).
34. Y. C. Loh, Internal stress gauges for cementitious materials, *Proc. Soc. Exp. Stress Anal.*, **11** (2), 13–28 (1954).

V

OTHER ELECTRICAL STRAIN AND DEFLECTION GAUGES

In Chapter IV it was explained how strain could be measured by its effect on the electrical resistance of a suitable gauge element. Strain can also be made to alter other electrical circuit parameters, such as capacity and inductance, by suitably designed transducers. Such instruments have the convenience of electrical measurement but may require more complicated equipment than is necessary for resistance gauges; their main advantage lies in their sensitivity and stability.

CAPACITY GAUGES

The possibility obviously exists of arranging a condenser in such a way that the spacing of its plates, and thus its capacity, is altered by variation of the distance between two gauge points. This principle has in fact been explored and has been used for strain measurement[1]. The simplest arrangement is suggested in Fig. 5.1 where two condenser plates are attached to the test piece by suitable posts mounted at some convenient distance apart; one of these plates is earthed and the other is connected to a circuit, by means of which the small change of capacity due to relative movement of the plates can be measured.

The simplest arrangement is that shown in Fig. 5.2 in which the measuring condenser C_1 and a shunt condenser C_2 are charged to about 100 V by a battery B_1.

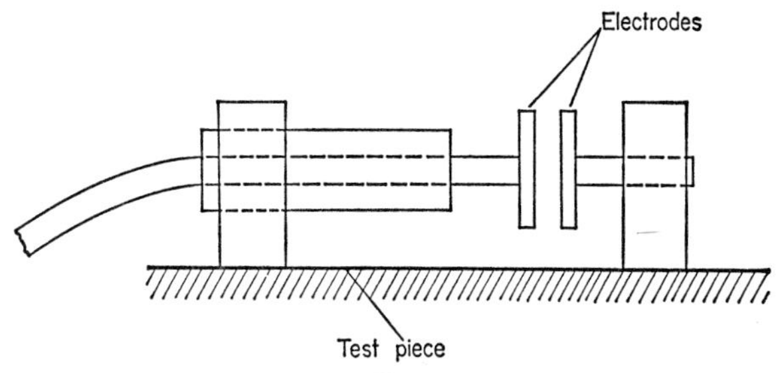

Fig. 5.1. Simple capacity gauge.

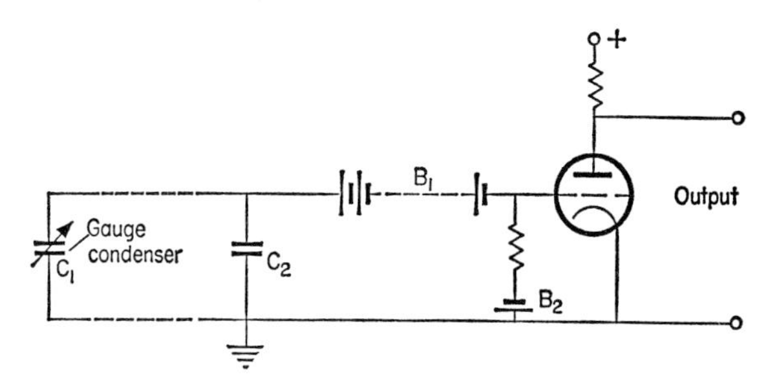

Fig. 5.2. Capacity gauge circuit; d.c. polarised.

The small grid bias battery B_2 sets the grid voltage at such a value as will ensure that the output from the valve is directly proportional to the charge in C_1, at least over a useful range. The output is fed to a cathode ray oscilloscope for observation or recording. The electrical characteristics of this circuit are such that it is unsuitable for measuring static or slowly varying strains; it is, however, satisfactory for observations of short period transient phenomena.

An alternative procedure, which makes it possible to measure static strains, is to apply a high frequency alternating current to the gauge and to measure the electrical effects of alterations in the gauge capacity. This can be done either by measuring the change in amplitude of this current or by making the gauge form part of the capacity in an oscillator circuit. The effect of this will be that as the gauge capacity changes so also will

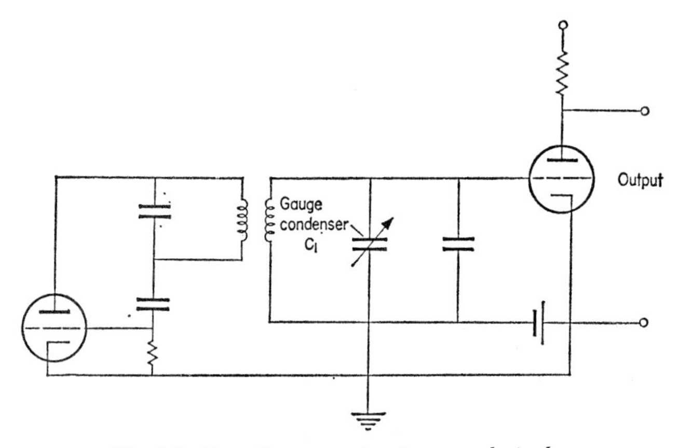

Fig. 5.3. Capacity gauge circuit; a.c. polarised.

the resonant frequency of the oscillator. The electrical arrangements for the first of these alternatives is somewhat similar to that shown in Fig. 5.1, except that the battery B_1 is omitted and the gauge is supplied from a fixed frequency oscillator through coupled inductances, as shown in Fig. 5.3. Considerable amplification of the output signal is required. The frequency modulated system is somewhat more complicated and makes use of a bridge circuit which in effect converts the frequency change into an amplitude change. This can be achieved by means of two resonant circuits arranged as shown in Fig. 5.4; when the two sides of the bridge are identical, the voltage e is zero, but any change in C_1, which is the gauge condenser, results in an out of balance voltage which can be amplified and displayed on a cathode ray oscilloscope or read on a suitable meter. By correct design a linear relationship

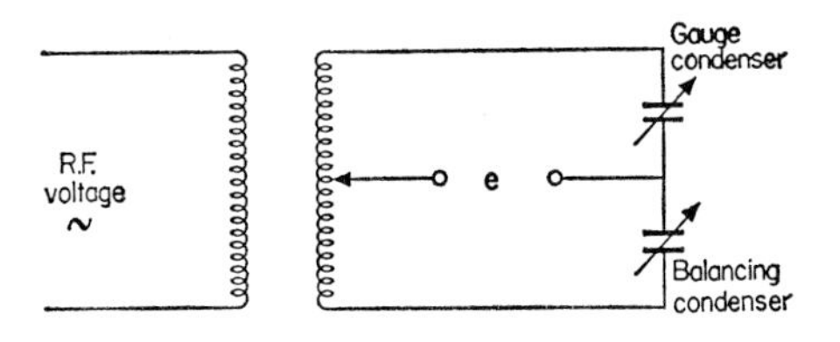

Fig. 5.4. Inductance–capacity bridge for capacity gauge.

can be obtained over a useful range between input frequency and output voltage. This principle is used in commercially produced instruments, such as the Fielden Proximity Meter, which can be used for strain measurement and for many other applications.

The capacitance of a two plate condenser in micro-farads is:

$$C = \frac{0{\cdot}009AK}{d}$$

where A is the area of the individual plates in square millimetres;

K is the dielectric constant of the material separating the plates;

d is the distance separating the plates in millimetres.

For example: a condenser having two plates each of area one square millimetre and a spacing between them of 2·5 mm will have a capacitance of $2\mu\mu$F. An instrument like the Fielden will discriminate a change of $0{\cdot}001\mu\mu$ F and thus an alteration of 50×10^{-6} mm. If the plate spacing was reduced to 0·25 mm, with the area kept constant, a change of 5×10^{-6} in the spacing is easily measured.

As in the case of resistance strain gauges, transducers can be made up for measuring pressure, displacement and other quantities. Small capacity elements have been used for strain measurement but have not been widely adopted presumably because their advantages do not outweigh the greater complication of the gauge itself and associated instrumentation, as compared with resistance strain gauges. On the other hand, capacity devices need not impose any restraint on the deformation of a specimen and may thus be extremely useful in measuring strain or displacements in easily deformed materials to which a resistance gauge could not be attached. Another valuable application of the capacity gauge is in the field of vibration studies where displacements of light objects can be measured without the need for any contact between the gauge and the object. Finally, capacity gauges may be devised to give a linear response up to large strains and at high temperature. In all cases, great care is necessary in the design of circuits associated with capacity gauges and in the selection of components and leads. This is in fact a task which is unlikely to be attempted by the novice although the use of commercial equipment is quite straightforward.

INDUCTANCE GAUGES

The third electrical circuit parameter which can be altered by mechanical means is the inductance of a coil [3,4]. This can be brought about in several ways, as indicated in Fig. 5.5. The most obvious possibilities are to vary the distance of a piece of soft iron from the poles of a electromagnet (i.e. the variable air gap arrangement of Fig. 5.5(a)), or to vary the position of the iron core in a coil, as in Fig. 5.5(b).

The eddy current method depends on the fact that the impedance of a coil is affected by moving a non-magnetic conductor through its field. Finally, use may be made of the fact that the magnetic characteristics of certain materials are affected by stress so that it is possible to use the magnetostriction effect to measure a force, as suggested in Fig. 5.5(d). The applications of this method are, however, rather limited.

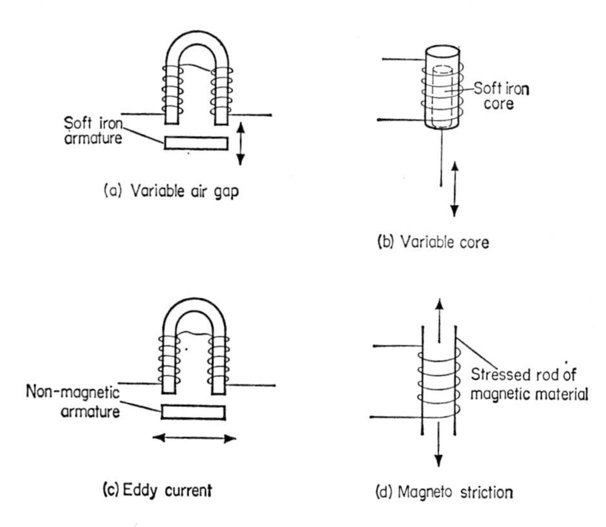

Fig. 5.5. Inductance gauge principles.

The simplest method of detecting the change of inductance of a coil is to place an ammeter in the supply line to the coil as in Fig. 5.6. This, however, has the serious disadvantage that the alteration of the current passing through the coil due to a

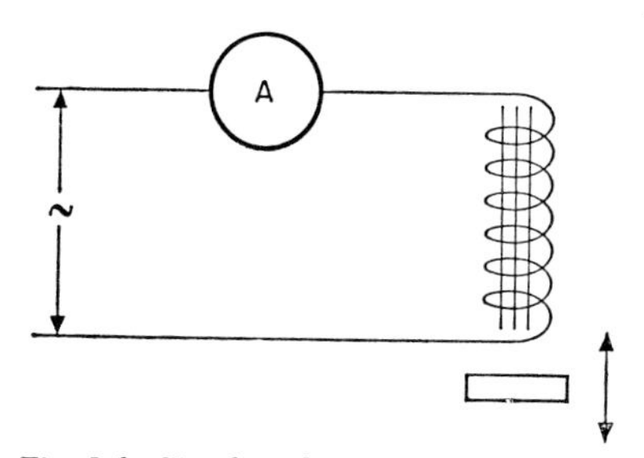

Fig. 5.6. Simple inductance gauge circuit.

change in the position of the armature (or the core in the case of a variable core instrument), may be small compared with the total current passing and thus only a small part of the ammeter scale is usefully employed. It is, therefore better to use a bridge circuit or an arrangement of inductances which limits the measured signal to the change produced by the effect being observed. The basic form of bridge circuit consists of four inductances arranged as in Fig. 5.7; it is clear that by suitable

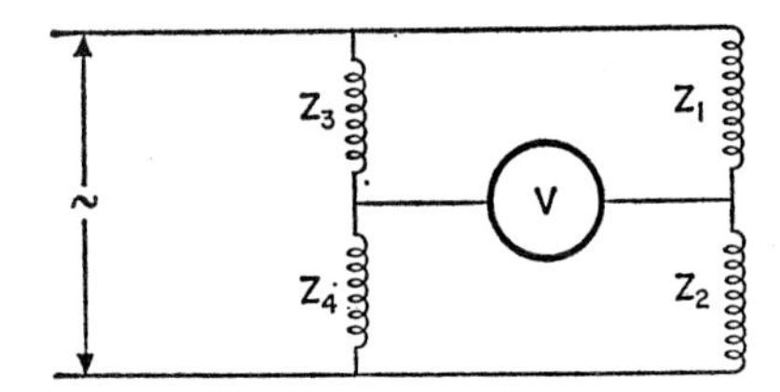

Fig. 5.7. Inductance gauge bridge.

selection of the values of the inductances it will be possible to obtain a zero reading on the voltmeter. Any subsequent alteration in, say, Z, will produce an out of balance voltage at V. As in a resistance bridge it is possible to make one, two or four of the bridge arms active in suitable cases. Similarly if Z_2 is the same mechanical form as Z_1 and is kept under the same ambient conditions it will provide temperature compensation in the event of alterations of inductance due to heating of the coils.

An alternative is to use three coils arranged as in Fig. 5.8. This device is known as a variable core differential transformer[5, 6]; when the core is symmetrically placed between the two identical secondary windings, equal and opposite voltages are induced in them. If the core is displaced, different voltages are induced in the two coils and a signal equal to the difference of these voltages is obtained. The output is proportional to the displacement of the core over a wide range. A substantial power output can be obtained from a differential transformer and sensitivity can be made to suit particular requirements by selecting a unit of appropriate physical size—the smaller the unit the greater the sensitivity obtainable.

Power for inductance gauges may be derived from the mains at 50 Hz for low frequency or static measurements. For

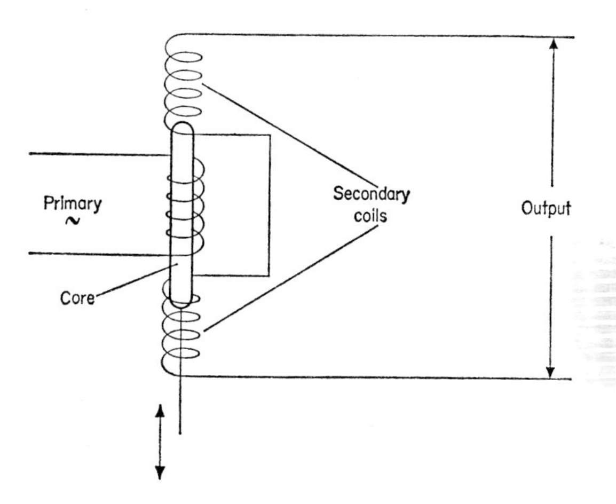

Fig. 5.8. Differential transformer.

measurements at higher frequencies the gauges must be supplied with a.c. at about ten times the frequency of the measured phenomenon; this can be derived from an oscillator and power amplifier. Because of iron losses in inductance gauges carrier frequencies are limited to about 5000 Hz.

It is not usually necessary to amplify the signal from an inductance gauge because of the high energy level obtainable except in the case of recording instruments with servo-operated, self-balancing potentiometers. The selection of a measuring or recording instrument for a particular application will be largely influenced by the frequency of the strain or displacement which is being observed. For example, pointer instruments are effective up to about 1 Hz and high speed recording galvanometers up to about 5000 Hz. Cathode ray oscilloscopes can of course be used over the whole effective range of inductance devices.

The main advantages of inductance gauges for strain measurement are:

(*a*) high stability
(*b*) high energy of the output signal
(*c*) high sensitivity; and
(*d*) robust mechanical construction.

To some extent (*a*) and (*b*) are inter-related, since instability arises in low energy systems from causes such as leakage currents due to moisture and pick-up from stray fields which are unimportant in inductance gauge circuits. A magnification factor between the strain being measured and the pointer movement of 10^4 is easily obtained with ordinary equipment which compares favourably with other types of gauges. Inductance gauges are obviously heavier and bulkier than, say, resistance gauges, but against this, there is frequently no need for amplifiers or delicate galvanometers. As previously mentioned, precautions are necessary against temperature effects. Self-heating of the coils is minimised by ensuring that their reactance is high compared with their resistance. Where single coil devices are used a compensating gauge, kept at the same temperature as the measuring gauge, should be included in the circuit.

VIBRATING WIRE GAUGES

The vibrating wire or acoustic strain gauge has been in use since about 1930[7, 8]. Its mode of operation is based on the fact that a stretched wire vibrates at a particular frequency which depends, amongst other things, on the tension in the wire. If the wire is stretched between two posts attached to a test specimen in which it is desired to measure the strain, as shown in Fig. 5.9, it is evident that any change in the distance between the posts due to strain will result in an alteration in the tension of the gauge wire, and thus in its natural frequency. In order to measure the frequency of vibration of the gauge wire, it is made of magnetic material and is caused to vibrate in the field of a permanent magnet surrounded by a coil by passing a pulse of d.c. current through the coil. A small

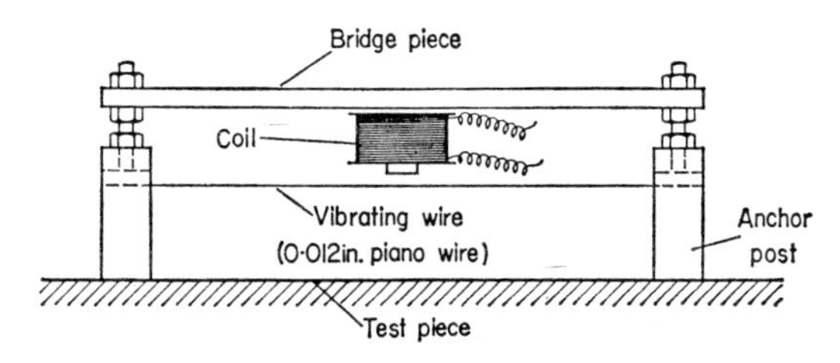

Fig. 5.9. Vibrating wire gauge.

alternating voltage is thus set up in the coil which, after amplification, is compared with a signal of similar amplitude by feeding one of the signals to be compared to the X plates of a cathode ray oscilloscope and the other signal to the Y plates. The frequency of the reference signal is adjusted until a circular or elliptical figure is observed on the screen; the two signals then have the same frequency which can be found if the source of the reference signal has been previously calibrated. The gauge wire is caused to vibrate by passing a pulse through the coil; the resulting vibration persists for a few seconds and by repeated plucking, the gauge and reference signals can be matched on the cathode ray tube. A block diagram of the apparatus is shown in Fig. 5.10.

In early versions of this equipment the reference signal was provided by a vibrating wire similar to the measuring gauge except that its tension could be altered by means of a micrometer screw and the signals were compared aurally using earphones. This is possible because as the two signals approach synchronism, the well known "heterodyne beat" is heard and the reference signal is adjusted until this just disappears. It is now common practice to replace the reference gauge wire with a variable oscillator[9], the calibration of which can be checked at any time by applying a signal from a standard tuning fork in place of the measuring gauge signal.

Any number of measuring gauges can be handled by incorporating a multi-way selector switch at the measuring gauge input terminals.

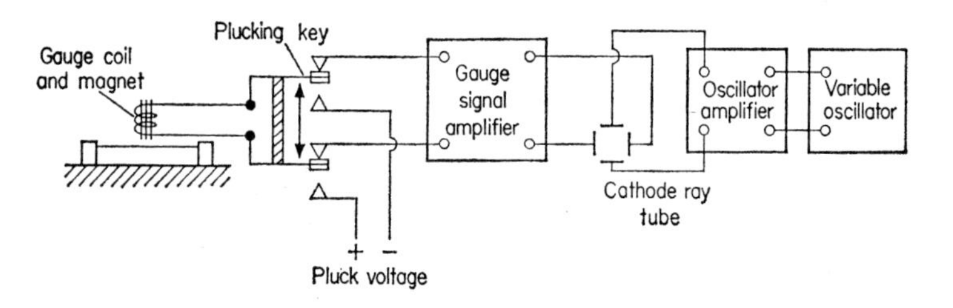

Fig. 5.10. Vibrating wire gauge; block diagram.

Vibrating wire strain gauges are marketed by Maihak A.G., Hamburg, and other manufacturers, and although these instruments are in very convenient form, a laboratory oscilloscope and oscillator can be used when a plucking circuit is added. It should be noted, however, that the oscilloscope will require to have a high gain amplifier in order to deal with the rather weak signal from the gauge. The oscillator should have a range of from 200 to 2000 Hz and should be able to read frequencies to 0·5 Hz. This requires a number of ranges and a slow motion tuning dial on the oscillator. These requirements tend to make the operation of the instrument rather slow and the instrument shown in Fig. 5.11 is fitted with a frequency meter which indicates the range to be searched for balance. A more elaborate method is to employ a digital frequency meter which will measure the frequency of the signal with the necessary accuracy instantly. A commercial instrument of this category is the Maihak MDS 4. This instrument can be made to record readings of a number of gauges on an electrically operated printer.

The frequency of vibration of a stretched wire is given by:

$$f = \frac{1}{2L}\sqrt{\left(\frac{T}{\rho}\right)}$$

where f is the frequency of vibration in c/s;
L is the length of wire;
T is the tension;
ρ is the density of material.

From this it is easily seen that the strain in the wire in terms of the frequency of vibration is:

$$\frac{\delta L}{L} = \frac{4L^2\rho}{AE} \cdot f^2$$

It is convenient to work in terms of strain units of $(f^2/1000)$

Fig. 5.11. Vibrating wire strain gauge; instrument.

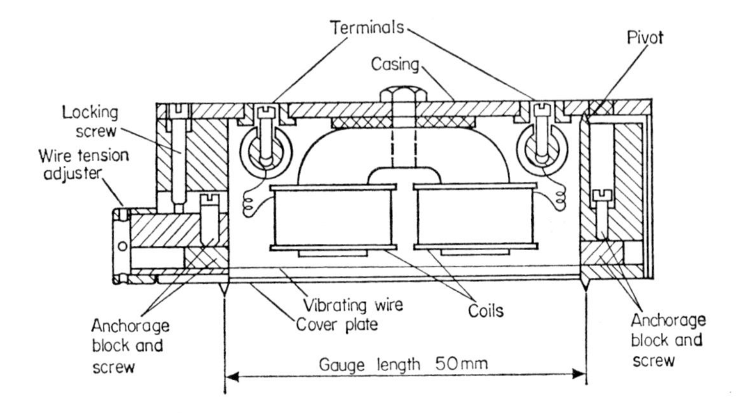

Fig. 5.12. Maihak MDS 14 demountable vibrating wire strain gauge. (Courtesy of Maihak A.G., Hamburg.)

and the strain in the material of the test piece will be obtained as the difference of two such numbers which relate respectively, to the initial and final frequencies of the gauge wire. Calibration can be carried out by means of a steel tension bar on which a vibrating wire is mounted, as shown in Fig. 5.11.

The accuracy of a vibrating wire strain gauge is of course determined by the accuracy of the electrical equipment used to measure the frequency of vibration. A sensitivity of $0{\cdot}5 \times 10^{-6}$ is quite easily achieved and a high degree of stability is attainable. It will be noted, for example, that as the measurement is of frequency the resistance of gauge leads and switch contacts does not come into the picture.

Gauges are of comparatively simple mechanical construction, robust and can be relatively inexpensive [11, 12]. Gauge lengths between 25 mm and 350 mm have been successfully used. The vibrating wire principle has been applied to a wide variety of gauges for the measurement of strain, pressure, deflection and temperature. Strain gauges can be demountable (see Fig. 5.12) or the wire may be mounted between two posts directly attached to the surface of the test piece as shown in Fig. 5.13.

The gauge wire is easily protected by a cover of design appropriate to the conditions: heavy cast-iron covers with rubber sealing gaskets have been used for gauges working under water.

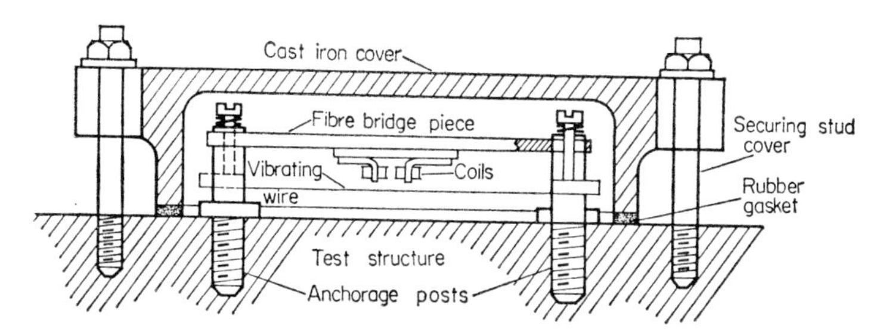

Fig. 5.13. Vibrating wire gauge mounted on steel structure with cast iron waterproof cover.

Gauges for internal strain measurements in concrete are made by Maihak and can be made up fairly simply in the manner shown in Fig. 5.14 (10).

Pressure measuring gauges can be made by attaching one end of the vibrating wire to a suitable diaphragm the deflection

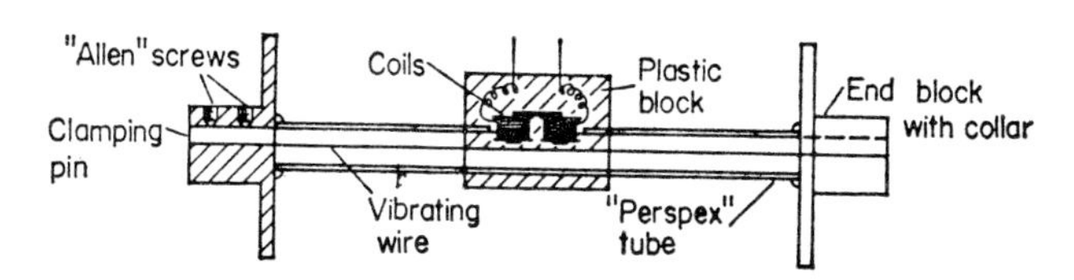

Fig. 5.14. Vibrating wire gauge for embedding in concrete.

of which due to pressure results in an alteration of the tension in the vibrating wire. Temperature gauges have a vibrating wire stretched inside a tube enclosed in an outer casing and free to move within it. As the ambient temperature changes, the length of the vibrating wire and the tension in it change;

it is, however, necessary for the inner tube to be of a material having a different coefficient of expansion from the gauge wire for the device to work.

Vibrating wire gauges are very suitable for applications requiring long term stability and high sensitivity and have therefore been used extensively in full scale tests on civil engineering structures such as tunnels, dams and earthworks.

BIBLIOGRAPHY

1. B. L. Carter *et al.*, Measurement of displacement and strain by capacity methods, *Proc. Inst. Mech. E.*, **152**, 215–21 (1945).
2. C. Brookes-Smith and J. A. Colls, Measurement of pressure, movement, acceleration and other mechanical quantities by electrostatic systems, *J. Sci. Instrum.*, **14**, 361–66 (1939).
3. B. F. Langer, Design applications of a magnetic strain gauge, *Proc. Soc. Exp. Stress Anal.*, **1** (2), 82–89 (1943).
4. W. H. Pickering, Reluctance gauges for telemetering strain data, *Proc. Soc. Exp. Stress Anal.*, **4** (2), 75–77 (1946).
5. H. Schaevitz, The linear variable differential transformer, *Proc. Soc. Exp. Stress Anal.*, **4** (2), 79–80 (1946).
6. A. Boggis, Design of differential transformer displacement gauges, *Proc. Soc. Exp. Stress Anal.*, **9** (2), 171–184 (1952).
7. N. Davidenkoff, The vibrating-wire method of measuring deformation, *Proc. ASTM*, **34** (2), 847–60 (1934).
8. R. S. Jerrett, The acoustic strain gauge, *Sci. Instrum.*, **22** (2), 29–34 (1945).
9. W. H. Ward and J. E. Cheney, Oscillator measuring equipment for vibrating-wire gauges, *J. Sci. Instrum.*, **37**, 88–92 (1960).
10. F. P. Potocki, Vibrating-wire strain gauge for long term internal measurements in concrete, *Engineer*, **206** (5369), 964–967 (1958).
11. T. K. Chaplin, Vibrating wire gauges and their uses, *Res. and Dev.*, **19**, 29–31 (March 1963).
12. J. C. Chapman, Stud welded vibrating wire strain gauges, *Engineer*, **206** (5361), 640–1 (1958).
13. I. K. Lee, The application of the vibrating wire measuring principle to the measurement of earth pressure, *Austral. J. App. Sci.*, **11** (1), 65–79 (1960).
14. F. Aughtie, The electrical measurement of strain, *Engineer*, **188** (4899), 713–4 (1949).
15. J. C. Simmons, A variable air gap inductive strain gauge, *Mag. Conc. Res.*, **6** (19), 31–34 (1955).
16. A. Ripperger, A piezoelectric strain gauge, *Proc. Soc. Exp. Stress Anal.*, **12** (1), 117–124 (1954).

VI

THE CALCULATION OF STRESSES FROM STRAINS

IN the preceding three chapters, the principles of various types of strain gauges have been discussed. Having made measurements of strain in an element, there remains the problem of deducing the stresses from these observations. The purpose of this chapter is to summarise briefly the methods of doing this in the case of elastic materials; it is assumed that the reader is already familiar with the elementary principles of stress, strain and elasticity. (References 21 and 22 give some information on the interpretation of strains measured above the elastic limit.)

PRINCIPAL STRESSES AND STRAINS

It is generally convenient to refer the stresses at a point in a body to three orthogonal axes, except in special cases where polar co-ordinates may be more appropriate. If we think of a small cubic element, the faces of which are parallel to a set of rectangular reference axes, these will in general be normal and shearing stresses on each of the cube faces. As is well known, however, it is possible to find a cube element at some inclination to the reference axes on the faces of which only normal stresses act; these are known as *principal* stresses. In exploring a stress system it is usually helpful to determine the magnitude and direction of the principal stresses at a sufficient number of points as a means of defining the stress conditions in the body.

In a field of non-uniform stress, the directions of the principal stresses vary from point to point. If a curve be drawn in such a way that the direction of one of the principal stresses is tangential to it, at every point along its length, then such a curve is termed a *line of principal stress* or *stress trajectory*. In a three dimensional stress system, lines of principal stress will form a system of three mutually orthogonal sets of curves; in a two dimensional system, the locus of points at which the principal stresses have the same inclination to a particular references axis is known as an *isoclinic line*; these lines will be referred to later in discussing the photo-elastic method of analysis.

Corresponding to the principal stresses are the *principal strains*. In an elastic body these are related to the principal stresses in the following way:

$$\epsilon_p = \frac{1}{E}[\sigma_p - \nu(\sigma_q + \sigma_r)] \tag{6.1}$$

$$\epsilon_q = \frac{1}{E}[\sigma_q - \nu(\sigma_p + \sigma_r)] \tag{6.2}$$

$$\epsilon_r = \frac{1}{E}[\sigma_r - \nu(\sigma_p + \sigma_q)] \tag{6.3}$$

where ϵ_p, ϵ_q, and ϵ_r, are the principal strains corresponding to the principal stresses σ_p, σ_q, and σ_r and ν Poisson's ratio.

PRINCIPAL STRAINS FROM MEASURED STRAINS

As a rule, the directions of the principal stresses will not be known in advance and if the principal strains are to be determined by measurement, it will be necessary to take strain measurements in as many arbitrary directions as will permit the calculation of the principal strains from the measured values. For example, in the two dimensional stress case there will be, in general, three strains related to the *X*- and *Y*-axes: normal strains ϵ_x and ϵ_y parallel to the axes and a shear strain γ_{xy}. If the strain is measured on any arbitrary line, say *AB* in Fig. 6.1, at an angle α to the *X*-axis, then the strain in this direction will be:

$$\epsilon_\alpha = \tfrac{1}{2}(\epsilon_x + \epsilon_y) + \tfrac{1}{2}(\epsilon_x - \epsilon_y)\cos 2\alpha + \frac{\gamma_{xy}}{2}\sin 2\alpha \qquad (6.4)$$

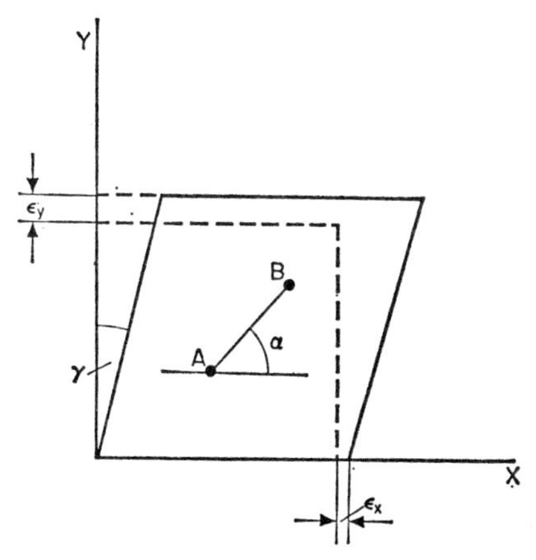

Fig. 6.1. Components of strain.

The principal strains will correspond to maximum and minimum values of ϵ_α so that on differentiating (6.4) and equating to zero, the directions of the principal strains ϵ_p and ϵ_q are found to be:

$$\tan 2\alpha = \frac{\gamma_{xy}}{\epsilon_x - \epsilon_y} \qquad (6.5)$$

Substituting (6.5) in (6.4) we then obtain for the principal strains:

$$\epsilon_{p,q} = \tfrac{1}{2}[\epsilon_x + \epsilon_y \pm \sqrt{\{(\epsilon_x - \epsilon_y)^2 + \gamma_{xy}^2\}}] \qquad (6.6)$$

Inspection of these equations will show that in order to find the principal strains it will be necessary and sufficient to measure the strains on three arbitrary gauge lines inclined at angles α_1, α_2, and α_3 to the X-axis. In principle, this will permit the calculation of ϵ_x, ϵ_y and θ_{xy}, and thence ϵ_p and ϵ_q. It is obviously advantageous to select values of α_1, α_2, and α_3 which will simplify the calculations, the most usual procedure being to take $\alpha_1 = 0°$, $\alpha_2 = 45°$, $\alpha_3 = 90°$ or alternatively $\alpha_1 = 0°$, $\alpha_2 = 60°$, $\alpha_3 = 120°$. These give rise to the gauge line dispositions shown in Figs. 6.2, referred to, respectively, as rectangular and equilateral (or delta) strain rosettes.

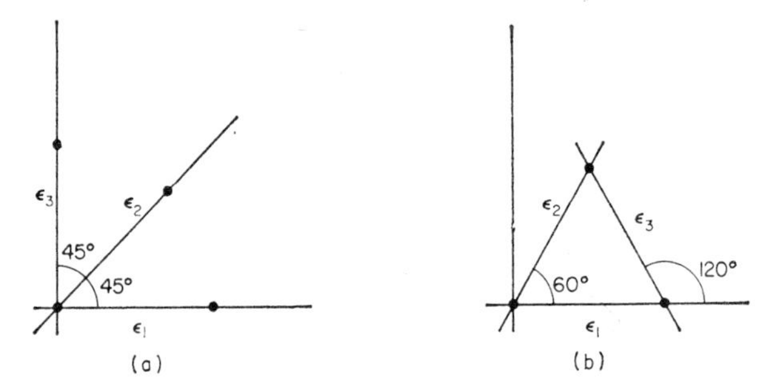

Fig. 6.2. Strain rosettes.

For a rectangular strain rosette it is easily shown that:

$$\epsilon_{p,q} = \frac{\epsilon_1 + \epsilon_3}{2} \pm \frac{\sqrt{2}}{2}\sqrt{[(\epsilon_1 - \epsilon_2)^2 + (\epsilon_2 - \epsilon_3)^2]} \quad (6.7)$$

$$\tan 2\alpha_{p,q} = \frac{2\epsilon_2 - (\epsilon_1 + \epsilon_3)}{\epsilon_1 - \epsilon_3} \quad (6.8)$$

For an equilateral or delta rosette:

$$\epsilon_{p,q} = \frac{\epsilon_1 + \epsilon_2 + \epsilon_3}{3} \pm \sqrt{\left[\left(\epsilon_1 - \frac{\epsilon_1 + \epsilon_2 + \epsilon_3}{3}\right)^2 + \right.}$$
$$\overline{\left. + \left(\frac{\epsilon_3 - \epsilon_2}{\sqrt{3}}\right)^2\right]} \quad (6.9)$$

$$\tan 2\alpha_{p,q} = \left[\frac{\frac{1}{\sqrt{3}}(\epsilon_3 - \epsilon_2)}{\epsilon_1 - \frac{\epsilon_1 + \epsilon_2 + \epsilon_3}{3}} \right] \tag{6.10}$$

Having calculated the principal strains the principal stresses are given by:

$$\sigma_p = \frac{E(\epsilon_p + \nu \cdot \epsilon_q)}{1 - \nu^2} \qquad \sigma_q = \frac{E(\epsilon_q + \nu \cdot \epsilon_p)}{1 - \nu^2} \tag{6.11}$$

If the directions of the principal stresses are known in advance, as for example on a line of symmetry, only two strain readings are necessary as the principal strains can be measured directly.

If a large number of readings are being taken the determination of principal stresses can become very tedious and therefore numerous devices have been proposed for speeding up the reduction of the strain readings. For comparatively small numbers of readings the construction of Mohr's circle provides a convenient method for the determination of the principal strains. The elements of the Mohr's strain circle are indicated in Fig. 6.3; it will be noted from this that the centre of the circle is at a distance $(\epsilon_p + \epsilon_q)/2$ from the origin and the radius of the circle is $(\epsilon_p - \epsilon_q)/2$. The state of strain on a gauge line at angle α is defined by drawing a radius at angle 2α to the horizontal axis, as is shown. The problem is thus to construct the circle from three measured strains such as ϵ_a. The construction shown in Fig. 6.4, suggested by Murphy[(1)], provides a neat solution. If ϵ_1, ϵ_2, and ϵ_3, are the measured strains on gauge lines at angles α_1, α_2, α_3 respectively to the principal stress direction, we proceed by setting out these strains to scale along a convenient horizontal line. Perpendiculars to this line are drawn through each of the three resulting points and at an arbitrary point, A, on the perpendicular through ϵ_2, lines are drawn at angles $(\alpha_2 - \alpha_1)$ and $(\alpha_3 - \alpha_2)$ to the perpendicular. The points of intersection of

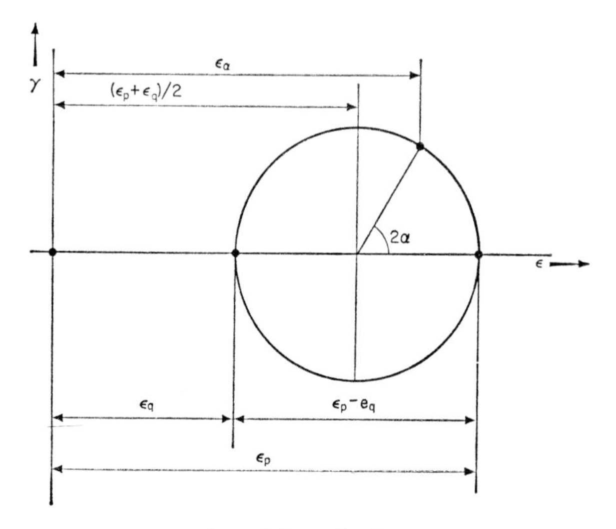

Fig. 6.3. Mohr's circle of strain.

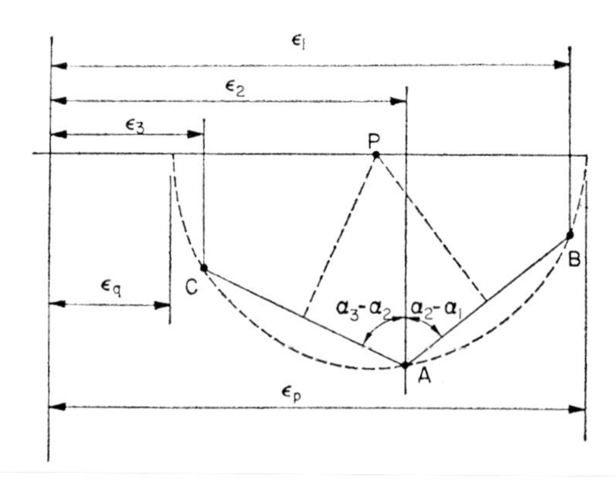

Fig. 6.4. Determination of principle strains by graphical construction.

these lines with the perpendiculars through ϵ_1 and ϵ_2 (B and C in Fig. 6.4) define points on the circumference of the strain circle, the centre of which is located by the intersection of the perpendicular bisectors of the two lines just drawn. The construction is completed by drawing the circle with centre P and radius $PA = PB = PC$.

This construction is applicable to any strain rosette and is particularly simple in the case of a rectangular rosette since in this case the centre is located by bisecting the line joining points C and B.

PRINCIPAL STRESSES

The principal stresses can now be found quite easily by drawing the Mohr's stress circle. The radius of the strain circle is $(\epsilon_p - \epsilon_q)/2$ and the radius of the required stress circle is $(\sigma_p - \sigma_q)/2$. The distances of the centres of the two circles from the origin are respectively $(\epsilon_p + \epsilon_q)/2$ and $(\sigma_p + \sigma_q)/2$. It is easily seen from equations (6.1) and (6.2) that

$$\sigma_p - \sigma_q = \frac{E}{1 + \nu}(\epsilon_p - \epsilon_q) \qquad (6.12)$$

and

$$\sigma_p + \sigma_q = \frac{E}{1 - \nu}(\epsilon_p + \epsilon_q) \qquad (6.13)$$

Thus, we must increase the radius of the strain circle by a factor $E/(1 + \nu)$ and the strain scale by a factor $E/(1 - \nu)$ to obtain the Mohr stress circle from which the magnitude and direction of the principal stresses and the maximum shearing stress can be found.

An alternative procedure is to draw a simple nomograph for the calculation of the principal stresses from the principal strains[10]. Naturally, a different diagram is required for each material, taking into account the appropriate values of Young's modulus and Poisson's ratio.

These simple procedures are quite adequate for dealing with readings from anything up to, say, 100 gauges, but more elaborate devices have been evolved for tests involving readings from hundreds or even thousands of gauges. One possibility is to carry out the reduction in one stage by means of a more complicated nomograph. Beyond that, it is possible, when using electrical strain gauges, to feed the gauge signals into a suitable form of computer or to employ gauge arrangements and circuits which will read out stress values directly. Descriptions of these devices will be found in the references given at the end of this chapter.

MAXIMUM SHEARING STRESS

Just as it is possible at any point in a three dimensional stress field to find three mutually perpendicular planes on which there are only normal stresses, so it is possible to find three planes on which the tangential or shear component of stress is at a maximum. These stresses are the maximum shearing stresses. In a two dimensional system the maximum shearing stress occurs on planes at 45° to the principal stress directions and is equal in magnitude to half the difference between the principal stresses. Lines of *maximum shearing stress* can be determined indicating the directions of the maximum shearing stress from point to point, analogous to the lines of principal stress.

Maximum shearing stress is of special significance in two contexts in experimental stress analysis. Firstly, in photoelasticity lines of equal principal stress difference and thus of maximum shearing stress are observed directly; they are known as *isochromatics* or *stress fringes*; secondly, the criterion of elastic failure of certain important materials (e.g. mild steel) is closely approximated by a limiting value of the maximum shearing stress which is, therefore, of particular interest in locating the onset and extent of plastic zones.

BIBLIOGRAPHY

1. G. MURPHY, A graphical method for the evaluation of principal strain from normal strains, *J. App. Mech.*, **12** (4), A-209–10 (1945).
2. W. M. MURRAY, An adjunct to the strain rosette, *Proc. Soc. Exp. Stress Anal.*, **1** (1), 128–133 (1943).
3. S. B. WILLIAMS, The dyadic gauge, *Proc. Soc. Exp. Stress Anal.*, **1** (2), 43–53 (1943).
4. E. E. HOSKINS and R. C. OLESON, An electrical computer for the evaluation of strain rosette data, *Proc. Soc. Exp. Stress Anal.*, **2** (1), 67–77 (1944).
5. J. H. MEIER and W. R. MEHOFFEY, Electronic computing apparatus for rectangular and equiangular strain rosettes, *Proc. Soc. Exp. Stress Anal.*, **2** (1), 78–101 (1944).
6. W. M. MURRAY, Machine solution of the strain rosette equations, *Proc. Soc. Exp. Stress Anal.*, **2** (1), 106–112 (1944).
7. R. BAUMBERGER and F. HINES, Practical reduction formulae for use on bonded wire strain gauges in two dimensional stress fields, *Proc. Soc. Exp. Stress Anal.*, **2** (1), 113–127 (1944).
8. J. H. MEIER, Improvements in rosette computer, *Proc. Soc. Exp. Stress Anal.*, **3** (2), 1–3 (1945).
9. K. J. BOSSART, A graphical method of rosette analysis, *Proc. Soc. Exp. Stress Anal.*, **4** (1), 1–8 (1946).
10. T. A. HEWSON, A nomographic solution to the strain rosette equations, *Proc. Soc. Exp. Stress Anal.*, **4** (1), 9–26 (1946).
11. N. GROSSMAN, A Nomographic rosette computer, *Proc. Soc. Exp. Stress Anal.*, **4** (1), 27–35 (1946).
12. F. A. MCCLINTOCK, On determining principal strains from strain rosettes with arbitrary angles, *Proc. Soc. Exp. Stress Anal.*, **9** (1), 209–210 (1952).
13. R. D. MINDLIN, The equiangular strain rosette, *Civ. Eng.*, **8** (8), 546–47 (1938).
14. S. S. MANSON, *An Automatic Electrical Analogue for* 45° *Strain Rosette Data*, Nat. Advisory Commit. for Aeronautics, Tech. Note 941. May 1944.
15. N. J. HOFF, A graphic resolution of strain, *J. App. Mech.*, **12** (4),. A1211–16 (1945).
16. A. H. WILLIS, The analysis of strain and its graphical representation, *Engineering*, **165** (4294), 457–60 (1948).
17. W. L. BRIDE, Solution of strain gauge rosettes, *Engineer* **207** (5374), 144–5 (1959).
18. S. ROGERS, Daisy: a system for acquiring, analysing and plotting strain gauge data, *Proc. Soc. Exp. Stress Anal.*, **17** (2), 87–90 (1960).
19. P. K. STEIN, A simplified method of obtaining principal stress information from strain gauge rosettes, *Proc. Soc. Exp. Stress Anal.*, **15** (2), 22–38 (1958).

20. ANON., Ultra simple method for analysis of strain rosettes. *R. and D*, **19**, 38–39 (1963).
21. K. SWAINGER, The measurement and interpretation of post-yield strains, *Proc. Soc. Exp. Stress Anal.*, **5** (2), 1–8 (1947).
22. J. TWYMAN, The interpretation into stresses of post-yield strains up to two per cent, *Proc. Soc. Exp. Stress Anal.*, **6** (1), 131–7 (1948).

VII

TWO-DIMENSIONAL PHOTO-ELASTICITY

THE methods and equipment described in earlier chapters have been concerned with the determination of stresses from measured strains or deflections. We come now to a completely different technique in which stresses are inferred from the optical behaviour of stressed transparent materials. This is the photo-elastic method in which a model of the structure or detail is made from a suitable transparent plastic and loaded in such a way as to simulate conditions in the prototype. It is possible to deduce the stress distribution in the stressed material from its appearance in polarised light. The fundamental phenomenon on which photo-elasticity is based was discovered by Sir David Brewster in 1816. Brewster was fully aware that it might be used as a method of stress measurement but as glass was the only material available during the nineteenth century for making models, little progress was made until the early decades of the twentieth century. This advance followed the introduction of celluloid for models and was largely due to the efforts of Coker and Filon whose *Treatise on Photo-elasticity* is still a much valued reference work in the subject [(1)]. Since the early 1930's new plastics suitable for photo-elasticity have become available and the method has been developed into a powerful and accurate stress analysis technique. As will be seen photo-elasticity has the advantage over most other methods of permitting the exploration of the whole of a stress field and not only at selected points. Furthermore, the method can be extended to the analysis of three dimensional systems.

OPTICAL THEORY

A light wave is an electro-magnetic disturbance, propagated in space or in transparent media. At any point the magnitude of the disturbance can be represented by a vector in a plane at right angles to the direction of propagation. Ordinarily, the direction of the light vector is quite random, but if its direction is at all times parallel to a fixed reference direction, the light is said to be plane polarised. Filters are available which have the property of transmitting only plane polarised light; the actual nature of these filters will be discussed in a later section of this chapter and in the meantime their existence may be taken for granted.

If two polarising filters are placed in series with their polarising axes at right angles to one another they constitute a plane polariscope, the first filter being termed the *polariser* and the second the *analyser*. With this arrangement, it will be noted that the analyser will not transmit any of the light from the polariser, so that on looking into the analyser, the field will be dark. Consider now the effect of introducing a strip of a suitable transparent material between the polariser and the analyser. If the strip is initially unstressed, no change will be introduced, but if the strip is subjected to a uniform tension, the light will, in general, be restored. This effect arises from the fact that, under stress the material acquires the property of breaking up the incident light into two components polarised in the directions of the principal stresses; the azimuths of polarisation of these components will thus differ by 90°. Further, the light waves in these beams are transmitted through the stressed plate with different velocities so that, when they leave it, the light waves in the two beams have acquired a relative path retardation which is found to depend on two factors, firstly, the difference of the principal stresses in the plate and, secondly, its thickness.

The analyser transmits only the components of those two beams parallel to its own plane of polarisation. The transmitted components combine to give a resultant light wave, but if the relative retardation is equal to the wavelength of the light used in the experiment optical interference results in the extinction of the transmitted light. If white light is used, the colour whose wavelength is equal to the relative retardation is extinguished and the light observed is coloured owing to the balance of the colours composing the white light being upset. Light whose wavelength is nearly equal to the relative retardation is partially extinguished. The action of the polariscope is illustrated in simplified terms in Fig. 7.1.

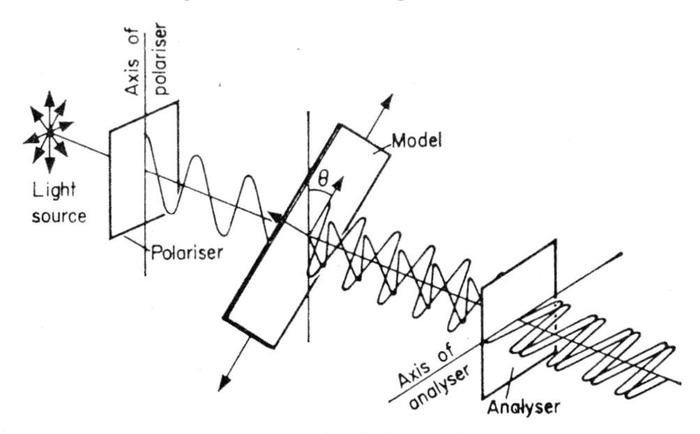

Fig. 7.1. Principle of plane polariscope.

The relative retardation (r) between the two transmitted waves is given by:

$$r = c(\sigma_p - \sigma_q)d \tag{7.1}$$

where σ_p and σ_q are the principal stresses, d the thickness of the plate and c a constant for the material, known as the stress-optical coefficient. Inspection will show that c has the dimensions of the reciprocal of stress; the unit employed is the brewster equal to 10^{-13} cm^2/dyne. In the uniformly stressed plate discussed above, the relative retardation is proportional to the tension in the plate; when viewed in white light, the sequence of colours observed as the stress is gradually increased as shown in Table 7.1. It will be noted that the colours are

Table 7.1

Colour Sequence in Stressed Transparent Material Viewed in White Light

Approximate Relative Retardation Å	*Colours Extinguished*	*Colour Observed*	*Order*
500	—	Grey	1st
2000	—	White	
4000	Violet	Yellow	
4500	Blue	Orange	
5000	Green	Red	
5900	Yellow	Purple	
6500	Orange	Blue	
7000	Red	Green	
8000	Deep red (1st) Violet (2nd)	Yellow	2nd
9000	Blue	Orange	
10,000	Green	Red	
11,800	Yellow (2nd) Violet (3rd)	Purple	
13,000	Orange–red (2nd) Indigo (3rd)	Emerald green	
14,000	Red (2nd) Blue (3rd)	Pale yellow	3rd
15,500	Deep red (2nd) Green (3rd) Violet (4th)	Pink	
17,000	Yellow (3rd) Indigo (4th)	Pale green	
19,000	Orange (3rd) Blue (4th)	White	
21,000	Red (3rd) Green (4th) Violet (5th)	Pale pink	4th
24,000	Deep red (3rd) Yellow (4th) Blue (5th) Violet (6th)	Pale green	
25,000	Orange (4th) Green (5th) Indigo (6th)	White	

As the retardation is increased, the overlapping of the extinguished colours becomes more and more pronounced; the colour sequence above the third order becomes a repetition of white, pink and green, getting paler each time until about the 7th order when the transmitted light is practically white.

extinguished in the order of the colours of the spectrum, beginning with those of shorter wavelength at the violet end. The change in the observed colour from red to blue corresponding to the extinction of yellow (wavelength 5893 Å) is fairly sharp and is of great value in photo-elastic work as the relative retardation, and thus the principal stress difference, can be inferred wherever this "tint of passage" is observed.

It will also be seen that the colour sequence is repeated with some modifications as the colours of the spectrum are extinguished a second time. The second, third and successive extinctions of a colour correspond to relative retardations of two, three and corresponding multiples of the wavelength of the light extinguished. Owing to the overlapping of the restored colours it is not possible to use white light where the principal stress difference is such that the relative retardation exceeds five or six wavelengths of yellow light.

If instead of white light, monochromatic light (i.e. light of one particular wavelength) is used, it will be appreciated that each time the relative retardation is equal to an exact multiple of the wavelength, the light will be extinguished, no matter how high the order may be.

So far, only the behaviour of a uniformly stressed plate has been considered; if the stress varies from point to point in the plate, the relative retardation and thus the colour observed will be the same at all points where $(\sigma_p - \sigma_q)$ is equal. The loci of such points in white light show the same colour and are thus known as *isochromatic lines*. In practice, the "tint of passage" isochromatics are quite sharply defined and can be traced for quantitative purposes. If monochromatic light is used the isochromatics are replaced by black lines referred to as "fringes", corresponding to a relative retardation equal to some multiple of the wavelength. Fig. 7.2 shows the appearance of these fringes in a beam with a central hole subjected to pure bending; in the portions of the beam at a distance from the hole, the fringes are equally spaced lines parallel to the neutral axis. Close to the hole, they form a pattern of curved lines.

As the fringes look alike, one must know the fringe value at some point in order to distinguish the fringe orders over the whole model. Examination of a model in white light is often very useful in this connection as, after a very little experience, it is possible to identify the various orders from their characteristic colours.

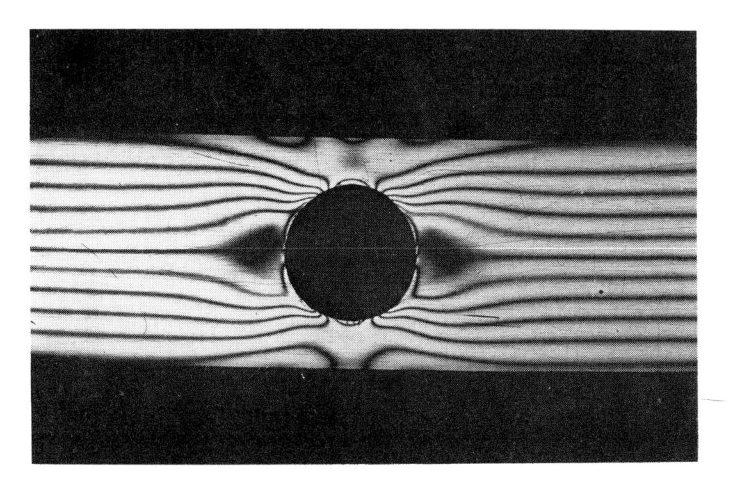

Fig. 7.2. Fringe photograph for straight beam with central hole.

If the direction of one of the principal stresses at a point in a stressed plate is parallel to the azimuth of polarisation of the incident light, the latter will pass through without being split up and will thus be stopped by the analyser. Thus in the case of a uniformly stressed strip, the field of the polariscope will go dark whenever the plane of polarisation of the incident light coincides with the axis of the strip. In a non-uniformly stressed plate the light will be extinguished whenever the direction of polarisation is the same as that of the principal stresses. The loci of points at which this occurs appears as a dark line known as an *isoclinic line*. By means of these lines, the directions of the principal stresses in a model can be determined. Wherever the direction of the principal stress is changing rapidly, the isoclinic lines are well defined, but in regions where the direction of the principal stress is nearly constant, the isoclinic lines will be found to be rather hazy. At any point where $(\sigma_p - \sigma_q) = 0$ (an *isotropic* or *singular* point), a dark point will be observed through which all isoclinic lines in the vicinity will pass. At such a point therefore the directions of the principal stresses are indeterminate.

From the foregoing, it will be concluded that the following information can be obtained:

(*a*) The direction of the principal stresses at every point in the model;

(*b*) The difference of the principal stresses at every point.

LINES OF PRINCIPAL STRESS

If a line is plotted such that the tangent to it is at every point in the direction of the principal stress, we have *a line of principal stress*. As in a two dimensional system, the principal stresses are at right angles to one another, the lines of principal stress form an orthogonal system of curves. These lines are very useful in visualising the stress system in a stressed plate and may be constructed from the isoclinic lines by the procedure indicated in Fig. 7.3. More accurate methods of drawing the lines of principal stress can be found in references 1–5.

DETERMINATION OF THE SEPARATE PRINCIPAL STRESSES

As the optical data afford a knowledge of the difference of the principal stresses, it is necessary to find some way of determining σ_p and σ_q separately. Many methods for this have been proposed, both by calculation and by subsidiary experiments. Of these only two will be described here—one experimental and the other numerical. Between them these methods are capable of dealing with most cases which are likely to be encountered. The student who wishes to study photo-elasticity in greater depth will, however, find it instructive to examine the various other methods described in references 21–25.

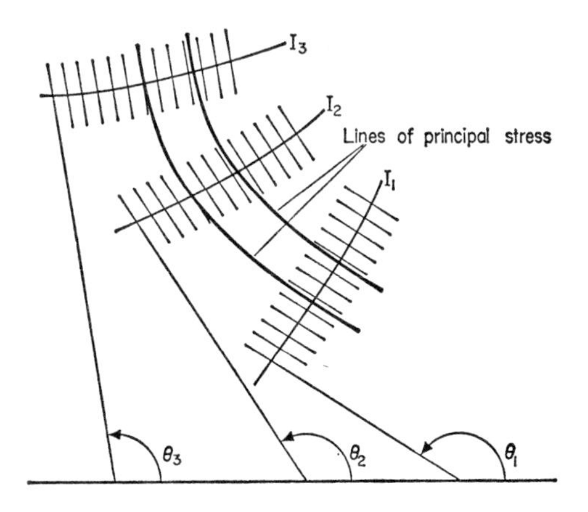

Fig. 7.3. Construction of lines of principle stress.

Drucker's Oblique-incidence Method[20]

In the above discussion of the photo-elastic effect, normal incidence of the light from the polariser on the model was assumed. By making certain observations with oblique incidence of the light beam, it is possible to obtain the separate principal stresses. As previously stated, the relative retardation is given by:

$$r = c(\sigma_p - \sigma_q)d \qquad (7.1)$$

If r is measured in terms of multiples of the wavelength of the light used and the stress-optical coefficient, c, is replaced by the material fringe value f, defined as the stress necessary to produce a first order fringe in a plate one inch thick, (7.1) can be re-written in the form:

$$n = \frac{(\sigma_p - \sigma_q)d}{f} \tag{7.2}$$

where n is the number of multiples of the wavelength—referred to as the fringe order. If now we put:

$$n_1 = \frac{\sigma_p d}{f} \quad \text{and} \quad n_2 = \frac{\sigma_q d}{f} \tag{7.3}$$

and substitute in (7.2) we have:

$$n = n_1 - n_2 \tag{7.4}$$

Suppose now that the model is rotated about the direction of the principal stress σ_p through an angle θ as indicated in Fig. 7.4(b). The principal stresses in the plane perpendicular

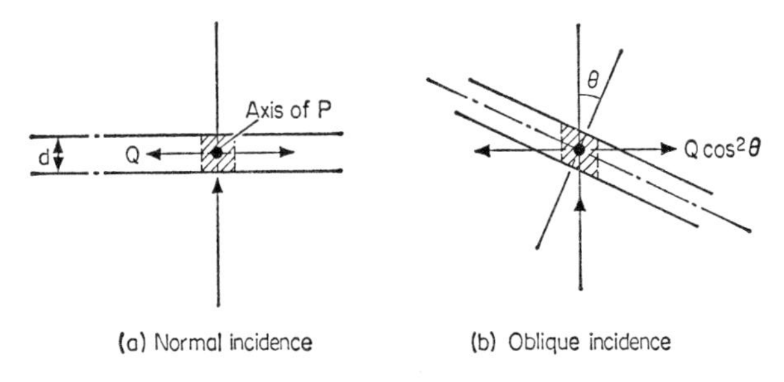

Fig. 7.4. Oblique incidence method.

to the direction of the light beam will then be σ_p, and $\sigma_q \cos^2 \theta$ and the light travels through a thickness $d/\cos \theta$ and the fringe order n_θ for oblique incidence will be:

$$n_\theta = \frac{\sigma_p - \sigma_q \cos^2 \theta}{f} \cdot \frac{h}{\cos \theta} = \frac{n_1 - n_2 \cos^2 \theta}{\cos \theta} \tag{7.5}$$

Solving (7.4) and (7.5) for n_1 and n_2:

$$n_1 = \frac{\cos \theta(n_\theta - n \cos \theta)}{\sin^2 \theta} \tag{7.6}$$

$$n_2 = \frac{n_\theta \cos \theta - n}{\sin^2 \theta} \tag{7.7}$$

σ_p and σ_q can then be obtained from (7.3). If the model is rotated about the direction of the principal stress σ_q corresponding relationships are obtained, namely:

$$n_1 = n - n_\theta^1 \cos\theta \tag{7.8}$$

$$n_2 = \frac{\cos\theta(n\cos\theta - n_\theta^1)}{\sin^2\theta} \tag{7.9}$$

where n^1 is the observed fringe order for oblique incidence in this direction.

This is a very convenient method for the determination of the separate principal stresses although practical difficulties may arise in the case of models requiring complicated loading arrangements. Such cases can, however, be dealt with by using the frozen stress technique which will be described in Chapter VIII. Some difficulty may arise also in thick models and at points where the stress gradient is very steep.

$(\sigma_p + \sigma_q)$ by Numerical Solution of the Laplace Equation(18, 19):

It is shown in textbooks on theory of elasticity that the sum of the principal stresses satisfies the Laplace equation:

$$\frac{\partial^2(\sigma_p + \sigma_q)}{\partial x^2} + \frac{\partial^2(\sigma_p + \sigma_q)}{\partial y^2} = 0 \tag{7.10}$$

If the boundary values of $(\sigma_p + \sigma_q)$ are known it is possible to solve this equation numerically using the finite difference formulae for (7.10) and an iterative procedure. The model is first covered by any convenient square network of reference lines and the values $(\sigma_p + \sigma_q)$ at the intersection of the grid lines with the boundaries of the model are found from the photo-elastic fringe pattern—this is possible because at a free boundary the principal stress normal to the boundary will be zero and thus the sum and difference of the principal stresses are numerically the same. It must be noted, however, that $(\sigma_p - \sigma_q)$ is always positive whereas $(\sigma_p + \sigma_q)$ is positive or negative according to which of the principal stresses is numerically the larger.

From the finite difference solution of the Laplace equation

it is known that the value of $(\sigma_p + \sigma_q)$ at a particular point is equal to the mean of that quantity at four equidistant nearby points. Thus referring to Fig. 7.5(a); if A, B, C and D are points on the reference grid equally spaced about O, and the sum of the principal stresses at these points is S_A, S_B, S_C, and S_D, then:

$$S_0 = \tfrac{1}{4}(S_A + S_B + S_C + S_D) \tag{7.11}$$

If A, B, C and D are not equidistant from O, but are disposed as shown in Fig. 7.5(b), then S_0 is given by:

$$S_0 = \frac{1}{1/ac + 1/bd}\left\{\frac{S_A}{a(a+c)} + \frac{S_B}{b(b+d)} + \frac{S_C}{c(c+a)} + \frac{S_D}{d(d+b)}\right\} \tag{7.12}$$

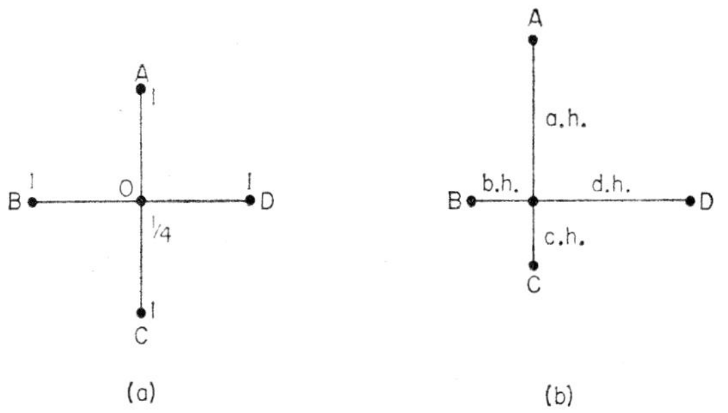

Fig. 7.5. (a) Finite difference pattern for Laplace equation $OA = OB = OC = OD = h$.
(b) Notation for incomplete squares.

The next step is to put trial values of $(\sigma_p + \sigma_q)$ at each of the interior grid points and then to apply (7.11) and (7.12) in a systematic manner so as to improve the initial values until eventually they satisfy the equations to within acceptable limits of accuracy. The nearer the initial values are to the correct ones, the more rapidly will the desired results be obtained, but any arbitrary set of initial values could be corrected to give the solution.

The method may be illustrated with reference to the example shown in Fig. 7.6.

The boundary values are written down, in this case in terms of fringe orders, and arbitrary values of $-1{\cdot}0$ and $-4{\cdot}5$ are taken at points $E.4$ and $E.8$, respectively. A first value of

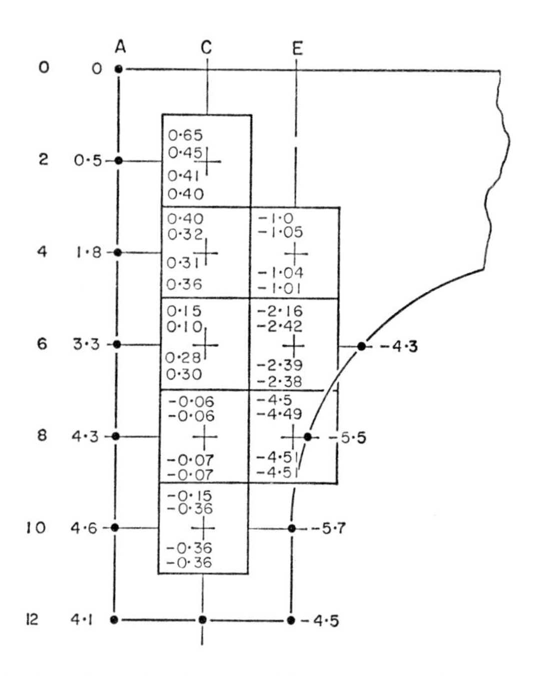

Fig. 7.6. Example of numerical determination of $(\sigma_p + \sigma_q)$.

$(\sigma_p + \sigma_q)$ at point $C.2$ is taken as the mean of the values at points $A.0$, $A.4$, $E.4$ and $E.0$, i.e.:

$$(\sigma_p + \sigma_q)_{C2} = \tfrac{1}{4}(0 + 1{\cdot}8 - 1{\cdot}0 + 1{\cdot}8) = 0{\cdot}65$$

Initial values at the other internal points are obtained in a similar manner.

These initial values are next improved by going round the grid systematically and adjusting the value at each grid point by putting it equal to the mean of the value at four adjoining points. Thus at point $C.2$, the first improved value is:

$$(\sigma_p + \sigma_q)_{C2} = \tfrac{1}{4}(0{\cdot}5 + 0{\cdot}4 + 0{\cdot}4 + 0{\cdot}5) = 0{\cdot}45$$

Then at $C.4$:

$$(\sigma_p + \sigma_q)_{C4} = \tfrac{1}{4}(0{\cdot}45 - 1{\cdot}0 + 0{\cdot}15 + 1{\cdot}8) = 0{\cdot}35$$

and so on round the network, equation (7.12) being applied, where necessary, at points near the boundary. This is repeated four times, by which time there is little alteration in the values.

The net shown in Fig. 7.6 is relatively coarse and more accurate values would be obtained by repeating the calculation using a net size half that of Fig. 7.6. The values of the initial coarse net are used to determine the starting values for the second net. It is sometimes necessary to use a still smaller net spacing in areas of the model where the stresses are changing rapidly.

The numerical method may be usefully combined with the oblique incidence method if the latter is used to find $(\sigma_p + \sigma_q)$ at key points in the grid. It may then be possible to dispense with an initial coarse grid.

PHOTO-ELASTIC APPARATUS

As previously mentioned, the basic optical apparatus in photo-elasticity is the polariscope consisting of two polarising filters and a light source; a suitable loading frame is of course also necessary and arrangements must be made for photographing the model in the apparatus.

Formerly, polarising filters were made by reflecting light from a glass plate at a particular angle as shown in Fig. 7.7(a). This provides an effective polarising filter and has the great advantage of cheapness and simplicity. It is, however, somewhat inconvenient and to permit the rotation of the plane of polarisation without changing the direction of the light beam, the triple reflection polariser [2] was devised on the principle shown in Fig. 7.7(b).

Natural crystals exhibit the property of double refraction by reason of which they transmit light polarised in two planes at right angles to one another; if one of these components

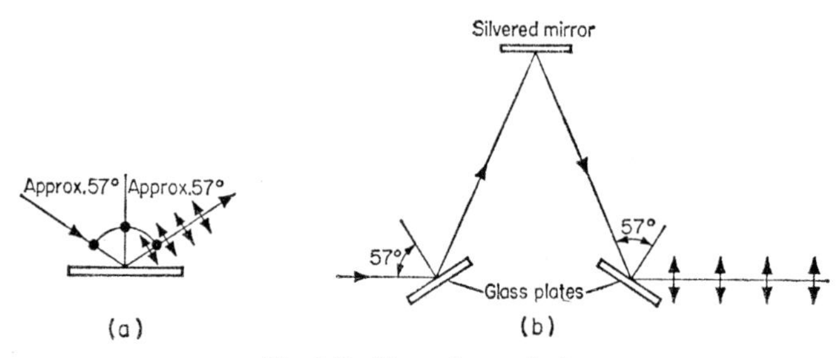

Fig. 7.7. Glass plate polarisers.

can be eliminated, the crystal will then serve as a polarising device. This can be done in certain cases by cutting the crystal and inserting a layer of material of suitable refractive index in such a way that one of the transmitted rays is reflected by total internal reflection whilst the other is transmitted through the device. In some crystals, however, one of the components is transmitted with only slight reduction in intensity whilst the other is reduced in intensity to the point of extinction. Such a crystal is said to be *dichroic*, and a synthetic substance, *herapathite*, possessing the characteristic, is used in the manufacture of Polaroid. This material consists of a sheet of cellulose nitrate or acetate in which herapathite crystals have been embedded, all oriented in the same direction, so that the whole sheet behaves as a very large dichroic crystal. Polaroid can be obtained in sheets up to 300 mm square and is very easily mounted on an optical bench to permit the axis of polarisation to be set in any required direction.

Polariscopes are usually provided with a pair of crystal plates known as "quarter wave plates", placed between the polariser and analyser. The first of these plates is set so that the plane polarised wave from the analyser is resolved into two equal plane polarised waves in directions at 45° to the axis of the polariser. On emerging from the quarter wave filter, these two components have a relative path retardation of quarter of a wavelength of a selected light wave and they combine to give a circularly polarised wave. The second quarter wave plate imparts a relative retardation of the same amount, but in the opposite sense, so that with no model between them they cancel each other out. If, however, there is a stressed model in the polariscope the circularly polarised light entering it will emerge elliptically polarised and after passing through the second quarter wave plate there will remain only two plane polarised waves with the same relative path retardation as in a plane polariscope, but as the circularly polarised light will not be affected by the direction of the principal stress axis in the model, no isoclinic lines will be visible. As the isoclinic lines tend to obscure the fringe pattern this arrangement has obvious advantages. As a rule the quarter wave plates are cut from mica plates which have been split down to the required thickness.

The lighting arrangement in a polariscope may be of two kinds: either an extended source without lenses or a concentrated source with lenses to give a parallel beam through the model. If it is desired to project an image of the model on a screen, the latter system is essential, but for photography either arrangement will give satisfactory results. For an extended white light source, tungsten lamps behind a frosted glass screen may be used whilst an ordinary bunched filament projection lamp provides a convenient concentrated source. Monochromatic light is conveniently obtained from a sodium vapour lamp.

Photographs can be taken either with a plate camera or with a 35 mm camera provided with extension collars. High contrast panchromatic plates or film must be used; care must be taken to get exposure and development times correct in order to reveal the structure of the fringes at points of stress concentration [11].

The straining frame for a polariscope should be made in such a way that any part of a model set up in it can be brought into the light beam from the polariser. This is usually achieved by mounting the frame on rails so that it can be moved transversely and by having an inner frame carrying the model and loading devices which can be moved vertically within this outer frame. Loading is very often by dead load and lever, but spring balances, air pressure bellows and other devices have been used to suit particular circumstances. As photo-elastic materials tend to creep under load, spring balances and similar devices require frequent adjustment to maintain a constant load on the model.

Photo-elastic apparatus suitable for general purposes is obtainable commercially, that shown in Fig. 7.8, produced by Sharples Engineering Company (Bamber Bridge) Limited, being an example of a polariscope fitted with 137 mm diameter Polaroid filters and alternative collimated or diffused lighting. Sodium and white light sources are instantly interchangeable.

MATERIALS FOR PHOTO-ELASTICITY

The essential properties of a transparent material for photo-elastic models may be summarised as follows:

(*a*) Good transparency
(*b*) Low material fringe value
(*c*) High elastic moduli and reasonable strength
(*d*) Low mechanical and optical creep
(*e*) Linear stress–strain and stress–retardation relation
(*f*) No initial double refraction
(*g*) Good machining properties
(*h*) Reasonable constancy of properties with variations of room temperature
(*i*) Moderate cost.

It is very difficult to find a material which is satisfactory in all these respects and the rate of development of photo-elastic

Fig. 7.8. Photo-elastic polariscope. (Courtesy of Sharples Engineering Co. Ltd., Preston.)

analysis has been largely set by the availability of model materials. Most of the early materials such as celluloid and various phenol formaldehyde resins, although still available, are now of historic interest and at present *epoxy* or *ethoxylene* resins provide the most satisfactory photo-elastic materials.

Epoxy resins suitable for photo-elasticity are available commercially in the United Kingdom as Araldite CT200 and Araldite MY753. Araldite CT200 is supplied in solid form which has to be melted down at 130–140°C and mixed with a liquid accelerator (Hardener 901) before casting. The proportion of accelerator recommended is 25–30 per cent by weight and the mixture should be cast at 120°C. Care has to be taken to eliminate air bubbles which may be entrained whilst the accelerator is being mixed with the resin or when casting. Metal moulds are the most suitable for accurate work and their surfaces must be coated with a suitable silicone mould release agent such as Releasil No. 7 or I.C.I. R.205. Thick metal moulds should be preheated to 130°C before casting. After pouring, castings must be cured at about 120°C for 16 hr.

Araldite MY753 is a liquid resin and is mixed with 8–10 per cent of Hardener HY951 at room temperature. Entrainment of air bubbles tends to be troublesome but may be minimised by heating the resin, *before* adding the hardener, to 30°C or more. If this is done, however, the mixture will have to be poured very soon after the addition to the hardener. It is important to note that considerable heat is generated by the reaction which takes place between the resin and the hardener; furthermore, the shrinkage on setting is considerable. This material is, therefore, not recommended by the manufacturers for large castings. In the author's experience, thickness of up to 25 mm can be successfully cast.

As the moulds for this material do not have to withstand high temperatures, plaster moulds may be used but again, for accuracy, metal ones are to be preferred. Plates can be cast between sheets of plate glass suitably spaced. Suitable mould release agents for Araldite MY753 are Araldite Mould Release QZ12 for room temperature curing or Releasil No. 7 or I.C.I. R.205 for curing at high temperatures.

The fringe value, elastic modulus and Poisson's ratio of Araldite CT200 and MY753 are shown in Table 7.2 along with the corresponding values for glass, celluloid and Catalin materials which have been used in the past for photo-elasticity. The development of photo-elastic materials can be followed by comparing the properties of the materials listed in this table.

Table 7.2

Properties of Photo-elastic Materials at Room Temperature

Material	*Fringe Value* $N/mm^2/mm$	*Elastic Modulus* N/mm^2	*Poisson's Ratio* $1/m$	*Tensile Strength* N/mm^2
Araldite CT200 (1952)	$16{\cdot}2 \times 10^{-3}$	$2{\cdot}9 \times 10^{3}$	0·35	72·3
Araldite MY753	$12{\cdot}4 \times 10^{-3}$	$3{\cdot}1 \times 10^{3}$	0·35	62·0
Glass (1816)	$285\text{–}875 \times 10^{-3}$	62×10^{3}	0·25	—
Celluloid (1900)	62×10^{-3}	$2{\cdot}1 \times 10^{3}$	0·4	48·2
Catalin (1940)	11×10^{-3}	$2{\cdot}1 \times 10^{3}$	0·42	48·2

The properties of plastics vary considerably according to their exact composition and are also affected by change of temperature. The values quoted in Table 7.2 should, therefore, be regarded as typical and it is essential to determine the fringe value of the material used in any particular experiment. The most convenient calibration specimen is a circular disc loaded in diametral compression. The principal stress difference at the centre of such a disc is:

$$\frac{8P}{\pi d \, . \, t}$$

where P is the load applied, d the diameter and t the thickness of the disc.

PREPARATION OF PHOTO-ELASTIC MODELS

The accuracy of photo-elastic results depends very largely on the careful preparation of the model and attention to detail in the actual test procedure. The first step in making a model from a plate of plastic material is to make sure that the plate is free from initial double refraction.

Having selected a stress free sheet of material, the surface is ground flat and polished. This may be done by laying a sheet of silicon carbide paper, as used by coach builders for cellulose finishing, on a glass plate and rubbing the plastic material over the abrasive with ample supply of water. A coarse grade of paper is used first to remove deep marks and to achieve a flat surface; thereafter the process is repeated on successively finer grades of paper until only scratches from the finest paper remain. Final polishing is carried out in a similar manner using a cloth and fine alumina instead of silicon carbide paper.

Models for qualitative examination may be prepared by scribing the outline on the surface of the material, cutting to within 1·5 mm by fretsaw or jigsaw, then filing carefully to the final dimensions. Greater accuracy is possible by clamping the blank between metal templates and then filing to size. This method is useful when a series of models has to be made differing only, say, in the size of a fillet; it ensures that the models are otherwise identical and that the edges are square. In general, models of the highest degree of accuracy are most conveniently prepared in a lathe or milling machine. Normal techniques are applicable, but very great care must be taken to avoid setting up machining stresses. The essential precautions are to use a very sharp cutting tool and to clamp the model by light pressure at many points.

It is useful to scribe a few very fine reference lines on the model during the machining operations as these will facilitate subsequent measurements on the fringe pattern.

Built up models are quite readily prepared as Araldite is

itself a cement. It is also possible to machine flanged models from thick plates to represent structural members and details.

TEST PROCEDURE

When the preparation of a model is completed, it is set up in the loading frame of the polariscope and adjusted so that it is accurately at right angles to the light beam. This can be checked by examining the appearance of the edges on the screen; if they appear as broad black lines, the edges of the model are not parallel to the light beam. If it proves impossible to get rid of a heavy black edge on all the boundaries at the same time the effect must be due to the edges of the model not being square to its surface or lack of parallelism of the light beam.

The model is next examined for machining stresses and other defects; if it is satisfactory in this respect, the load is applied. The development of the fringe pattern should be observed on the camera screen and sufficient notes taken to permit the identification of fringes in the monochromatic photograph. The model is then photographed, the time of taking the photograph after applying the load being noted. This latter point is important in eliminating errors due to optical creep by photographing a calibration specimen at the same time interval after loading as the model itself; this procedure was discussed in relation to mechanical creep in Chapter I.

After photographing the fringe pattern in monochromatic light, measurements may be taken of fractional fringe orders at any essential points by the Tardy method:

(1) Find the direction of the principal stress at the point of measurement by rotating the polariser and analyser together.

(2) Insert the quarter wave plates with their axes at 45° to that of the polariser and analyser.

(3) Rotate the analyser only until the nearest fringe to the point of measurement is displaced to give extinction at this point. The angle (θ) through which the analyser has been rotated is then observed.

(4) If the rotation of the analyser moves the nth order fringe in the direction of increasing fringe order to give extinction at the required point, then the fringe order there is:

$$\left(n + \frac{\theta}{180}\right)$$

If the rotation had been in the opposite direction the next higher fringe would have been moved to give extinction and the angle of rotation would have been $(180 - \theta)$ and the fringe order would then be:

$$\left(n - \frac{180 - \theta}{180}\right)$$

Oblique incidence observations should also be made at this stage. Finally, the isoclinic lines are plotted by projecting an image of the model on a glass screen behind which a sheet of tracing paper is mounted. White light will be found best for this operation for which the quarter wave plates are removed; the angle of the polariser and analyser is altered by five or ten degree steps and the isoclinics traced at each setting.

SOURCES OF ERROR IN PHOTO-ELASTICITY

Photo-elasticity requires a fair measure of experience and skill for accurate results so that it is well to be aware from the outset of some of the more important sources of error. These include:

(*a*) Initial double refraction in the model due to machining or other internal stresses.

(*b*) Lack of dimensional accuracy: the necessary accuracy of a model differs from one case to another and should be assessed in advance. For example, a 12 mm deep member stressed in bending would have to be accurate to ±0·025 mm to ensure accuracy of stresses to within 2 per cent. Accuracy of boundary stresses can also be lost if the edges of the model are not square.

(*c*) Faulty loading conditions may result from inaccurately placed loads, buckling, local plastic yielding or friction at supports. It is desirable that loads should be applied by dead load, with or without a lever system. If spring loading systems are used, care must be taken to observe that the load does not fall off as the result of strain creep.

(*d*) Optical creep has been mentioned already: modern plastics used for photo-elasticity are much less troublesome in this respect than older materials but care is still necessary.

(*e*) The properties of plastics are sensitive to quite small changes in temperature so that it is important to ensure that the calibration specimen is tested at the same temperature as the model.

BIBLIOGRAPHY

1. E. G. Coker and L. N. G. Filon, *A Treatise on Photo-elasticity*, 2nd Ed. Cambridge Univ. Press, 1957.
2. L. N. G. Filon, *A Manual of Photo-elasticity for engineers*. Cambridge Univ. Press, 1936.
3. M. M. Frocht, *Photoelasticity*, 2 vols. J. Wiley, New York, 1941 and 1946.
4. A. W. Hendry, *An Introduction to Photo-elastic Analysis*, Pergamon Press, to be published.
5. H. T. Jessop and F. C. Harris, *Photo-elasticity: Principles and Methods*. Cleaver Hume, London, 1949.
6. R. B. Heywood, *Designing by Photo-elasticity*. Chapman & Hall, London, 1952.
7. R. D. Mindlin, A review of the photoelastic method of stress analysis, *J. App. Phys.*, **10**, 222–47, 273–94 (1939).
8. R. E. Arthur, Introduction to the theory of photoelasticity, *J. Roy. Aero. Soc.*, **47**, 263–72 (1943).

9. A. F. C. Brown and V. M. Hickson, Photoelastic Laboratory at the National Physical Laboratory, *Engineering*, **171** (4454), 701–4 (1951).
10. H. T. Jessop, The optical system in photoelastic observations, *J. Sci. Instrum.*, **25** (4), 124–6 (1948).
11. J. H. Flanagan, Photoelastic photography, *Proc. Soc. Exp. Stress Anal.*, **15** (2), 1–10 (1958).
12. J. W. Dally and F. J. Akimag, Photographic method to sharpen and double isochromatic fringes, *Proc. Soc. Exp. Stress Anal.*, **19** (1), 170–5 (1961).
13. D. Post, Isochromatic fringe sharpening and fringe multiplication in photoelasticity, *Proc. Soc. Exp. Stress Anal.*, **20** (2), 143–156 (1962).
14. H. Spooner and C. D. McConnell, An ethoxylene resin for photoelastic work, *Brit. J. App. Phys.*, **4** (6), 181–4 (1953).
15. J. D'Agostino *et al.*, Photoelastic stress analysis in epoxy adhesives and casting resin as photoelastic plastics, *Proc. Soc. Exp. Stress Anal.*, **12** (2), 123–128 (1955).
16. R. S. Alwar, Araldite used for model analysis of composite structures, *J. Sci. Instrum.*, **12** (9), 526–7 (1961).
17. D. A. Senior, High speed photoelasticity: development and application, *Engineering*, **186** (4824), 248–253 (1958).
18. G. H. Shortley and R. Weller, Calculation of the stresses within the boundaries of photoelastic models, *Trans. A.S.M.E.*, **61** (1939).
19. G. H. Shortley and R. Weller, The numerical solution of Laplace's equation, *J. App. Phys.*, **9**, 334–48 (1938).
20. D. C. Drucker, The method of oblique incidence in photoelasticity, *Proc. Soc. Exp. Stress Anal.*, **8** (1), 51–66 (1950).
21. I. Brodie, A rapid method for separating the principal stresses in plane photoelasticity, *Brit. J. App. Phys.*, **9** (5), 201–3 (1958).
22. D. Vasarhelyi, Contribution to the calculation of stresses from photoelastic values, *Proc. Soc. Exp. Stress Anal.*, **9** (1), 27–34 (1951).
23. S. P. Christodoulides, A photoelastic method of two-dimensional separation of stresses along a line of symmetry by using the isochromatic fringes only, *J. App. Phys.*, **7** (5), 190–4 (1956).
24. R. K. Muller, A simple method for determining the principal stress trajectories in the plane photoelastic model test, *Beton u. Stahlbetonbau*, **53** (12), 307–9 (1958).
25. A. Pirard, Remarks on moiré method in photoelasticity, *Rev. Univ. Mines*, (16) 9 (4), 177–200 (1960).

VIII

THE FROZEN STRESS METHOD AND SURFACE COATING TECHNIQUES

An outline was given in Chapter VII of the two dimensional method of photo-elasticity which permits the exploration of the stress distribution in a plastic model cut from a flat plate. The present chapter discusses three useful stress analysis techniques, two of which are developments of the photo-elastic method. The first is the application of the "frozen stress" phenomenon to the analysis of two and three dimensional stress systems, the other two are surface coating techniques which permit the exploration of surface strains and stresses in elements of any degree of geometrical complication. In one of these methods the coating is a photo-elastic material and the optical effects brought about in it are examined in polarised light using a reflection polariscope. The remaining technique consists of applying a brittle coating which reveals the surface strains in the test piece by cracking.

THE FROZEN STRESS EFFECT

It was discovered by Solakian in 1935 that if a bar of a particular phenol formaldehyde plastic was stressed at 75°C and cooled to room temperature whilst still under load, fringes could be observed in it even after the removal of the load. This effect was investigated further by Hetényi [5] and others and it was demonstrated that, in any given case, the fringe pattern "frozen" into the material is the same as would be observed in a specimen loaded at room temperature although the loads required to produce it would be very much less. Furthermore, it was shown that the model could be cut into pieces without disturbing the fringe pattern in the various pieces. This is illustrated in Fig. 8.1 which shows two frozen stress models which have been cut through.

Fig. 8.1. Frozen stress fringe patterns.

The explanation of these rather remarkable phenomena is to be found in the molecular structure of the plastic material. This can be considered to consist of two phases, one of which is rigid at room temperature, becoming effectively fluid at high temperatures; the second phase retains its rigidity at all temperatures, but this is much lower than the low temperature rigidity of the first phase. The behaviour of the material can thus be represented by the simple mechanical model shown in Fig. 8.2; here the first phase is represented by a material which is solid at room temperature but which melts at some temperature around 40°C. The second phase is represented by a spring.

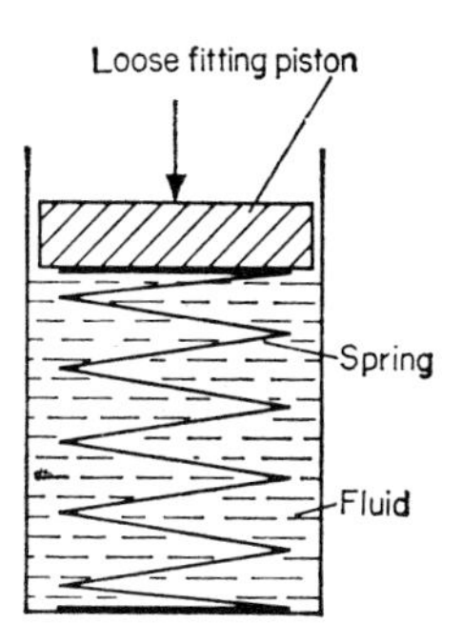

Fig. 8.2. Mechanical model of behaviour of plastics in frozen stress experiment.

At room temperature, imposed loads will be carried by the solid phase with relatively small deformation. If a load is applied at a temperature above the melting point of the first phase, it will be carried entirely by the second phase, represented by the spring. The deformations for a given load will in this case be greater than at room temperature. If the system is cooled down with the load still acting it will be seen that the spring will be "frozen" in its deformed state when the first phase again becomes solid and will remain deformed even when the load is removed. Furthermore, if the block of material is cut through, the spring will remain unaffected.

In plastics which exhibit this behaviour, the spring system takes the form of a strong polymerised network and is surrounded by a fusible phase having negligible rigidity above the softening temperature. The internal stresses in the network when it is frozen in a deformed state are balanced by small intermolecular forces in the fusible phase and it is thus possible to cut the material without upsetting the fringe pattern. The practical importance of this is that it permits the exploration of the internal stresses in a three dimensional structure by examining the photo-elastic fringe pattern in slices cut from a plastic model which has previously been subjected to a frozen stress loading cycle. The frozen stress method is also very useful in two dimensional experiments in which it is inconvenient to examine the model whilst it is actually under load, as, for example, if the stress system is due to centrifugal force.

PROPERTIES OF PLASTICS USED FOR FROZEN STRESS EXPERIMENTS

A number of plastics exhibit the frozen stress effect but as in the case of normal two dimensional work, epoxy resins are by far the best of those at present available. The relevant properties of Araldite CT200 and Araldite MY753 are tabulated below:

Table 8.1

Material	*Loading Temperature* °C	*Material Fringe Value* $N/mm^2/mm$	*Young's Modulus* N/mm^2
Araldite CT200	100	$0{\cdot}38 \times 10^{-3}$	13·4
Araldite MY753	80	$0{\cdot}37 \times 10^{-3}$	13·1

Comparing these values with the corresponding room temperature figures (of Table 7.2, p. 91), it will be observed that the fringe value is reduced by a factor of about 40 and Young's modulus by a factor of about 200. This means that the load necessary to produce a fringe pattern with a given maximum fringe value will be about one fortieth of that required at room temperature. The deformations will, however, be about five times as great and this may occasionally be rather troublesome because the strains developed may be large enough to alter the geometry of the model to a significant degree. A corollary of this is that it is frequently necessary to restrict the load in order to keep the deformations within acceptable limits and thus the number of fringes available for quantitative estimation tends to be lower than in two dimensional work.

PROCEDURE IN FROZEN STRESS ANALYSIS

The general procedure to be followed in carrying out a test by the frozen stress method will already be clear from the description of the effect and the information given in the preceding paragraph regarding materials. The model is first cast and machined to shape as necessary; it is then set up in a suitable loading rig in an oven and its temperature raised to about 100°C according to the material being used. The rate of heating and the period for which the model is kept at the maximum temperature depends on the sectional thickness of the model; gradual heating is necessary to avoid undue thermal stresses and sufficient time must be allowed at the maximum temperature to permit the achievement of uniform conditions through the thickness of the material. Similar considerations apply to the cooling of the model after the loads have been applied and a typical cycle might be: heating to 100°C—4 hr; temperature constant at 100°C—2 hr; cooling—4 hr.

The material fringe value should be determined at the same time as the model is tested. This is conveniently done by loading a disc specimen in diametral compression, as in two dimensional work. It will be possible to develop a sufficient number of fringes in the test specimen by loading with dead weight in a simple frame.

After the heating and cooling cycle is completed the model is removed from the oven and, if two dimensional, examined in the polariscope. In the case of a three dimensional model it will be necessary to cut slices in carefully selected directions for optical analysis. In qualitative experiments slices can be cut by hand or machine sawing, but in more accurate work slicing is best performed in a milling machine. Whatever the method of cutting, care must be taken to avoid the generation of excessive heat which would distort the fringe pattern in the vicinity of the cut. The thickness of slices depends to some extent on the configuration of the model but should be as small as possible consistent with the development of retardations measurable in the polariscope. After cutting, the surfaces of the slices may be too rough for the transmission of light but this difficulty is easily overcome by wetting with a fluid of the same index of refraction as the plastic.

INTERPRETATION OF FRINGE PATTERN IN SLICES

In the general case, there are five unknown quantities in a three dimensional stress system, namely the three principal stresses and two angles defining their directions to some axes of reference. It is possible to determine from photo-elastic observations the three principal stress differences at a point in a model and their orientation, but unless one of the principal stresses is known this information will not give the principal stresses themselves. Many ways of resolving this problem have been proposed but a discussion of these is beyond the scope of this book. On the other hand, the highest stresses are very often to be found at the surface of an element or in a plane of symmetry. In these cases a slice can be cut normal to the direction of one of the principal stresses and the fringe pattern observed in such a slice will give the difference of the other two. The separate principal stresses in such a slice are conveniently found by oblique incidence observations. Applied in this way to planes of symmetry and slices parallel to the surface of a model, the frozen stress technique can give results whose accuracy is comparable to that obtained by two dimensional photo-elasticity [8].

PHOTO-ELASTIC COATING TECHNIQUES

The idea of measuring strains photo-elastically by cementing transparent material to the surface of the test piece, is by no means new, but it is only in recent years that materials have become available which show measurable optical effects at the order of strain to be expected in a metal within the elastic range. For example, a simple tensile stress of 7 N/mm^2 in an aluminium specimen would result in a strain of about 100×10^{-6}; this in turn would correspond to a stress of approximately 0·3 N/mm^2 in Araldite. If the Araldite was 2·5 mm thick, this would produce about 0·2 fringe when viewed in a reflection polariscope. It is thus quite feasible to measure elastic stresses in metals in this way although the sensitivity will not be as high as in normal photo-elasticity.

The optical equipment for this technique comprises a simple reflection polariscope arranged as shown in Fig. 8.3(a). Reflection of the light beam from the surface of the test piece is achieved either by polishing it, or more easily by using an adhesive containing aluminium. The direction of the principal stresses can be found by removing the quarter wave plates and rotating the polariser and analyser together until an isoclinic line is brought over the point at which the measurement is being taken. The fringe order at the point can be determined by the Tardy method, described on p. 93, in relation to two dimensional photo-elasticity. This of course gives a measure of the maximum shear stress at the observed point. The separate principal stresses can be evaluated by the oblique incidence method using the optical arrangement suggested in Fig. 8.3(b). The method of calculation in both cases will be similar to that used in conventional two dimensional work; thus, for normal incidence the relative retardations:

$$r = c(\sigma_p - \sigma_q)2t \qquad \text{or} \qquad \frac{r}{2tc} = (\sigma_p - \sigma_q)_c \tag{8.1}$$

where $(\sigma_p - \sigma_q)_c$ denotes the difference of the principal stresses in the coating, c the stress optical coefficient and t the thickness of the coating. It will be noticed that the path distance through

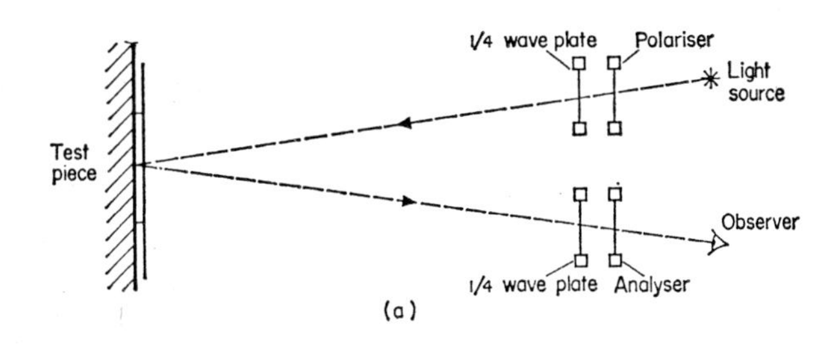

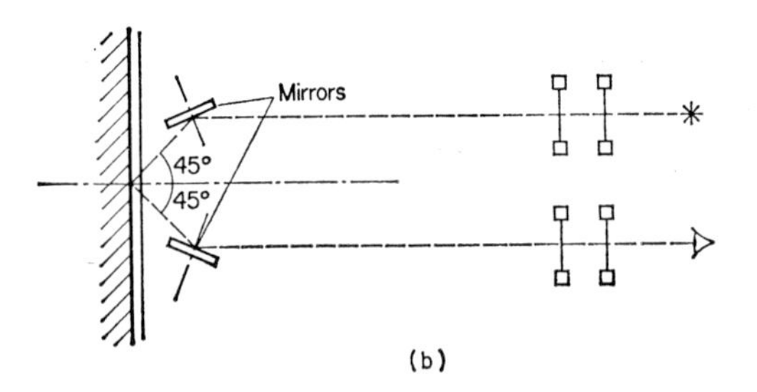

Fig. 8.3. Reflection polariscope for photo-elastic coating method ("Photostress" system).

the material is $2t$ as the light passes twice through the material. It will be remembered from equation (6.11) that:

$$\sigma_p = \frac{E}{1 - (1/m)^2}\left(e_p + \frac{1}{m} e_q\right)$$

$$\sigma_q = \frac{E}{1 - (1/m)^2}\left(e_q + \frac{1}{m} e_p\right)$$

Thus, considering the coating:

$$(\sigma_p - \sigma_q)_c = \frac{E_c}{1 - (1/m)^2}\left[\left(e_p + \frac{1}{m} . e_q\right) - \left(e_q + \frac{1}{m} e_p\right)\right]$$

$$= \frac{E_c}{1 + (1/m)_c} . (e_p - e_q) \quad (8.2)$$

Similarly in the material of the test piece:

$$(\sigma_p - \sigma_q)_m = \frac{E_m}{(1 + 1/m)_m} . (e_p - e_q) \quad (8.3)$$

From (8.1) and (8.2):

$$(e_p - e_q) = \frac{(1 + 1/m)_c}{E_c} . \frac{r}{2t . c} = \frac{r}{2t \cdot K} \quad (8.4)$$

where $K = \dfrac{C . E_c}{(1 + 1/m)_c}$

Substituting in (8.3) we obtain:

$$(\sigma_p - \sigma_q)_m = \frac{E_m}{(1 + 1/m)_m} . \frac{r}{2t . K} \quad (8.5)$$

This expression permits the calculation of the principal stress difference in the material of the test piece. K is a factor depending on the optical properties and elastic constants of the coating material and is found by a suitable calibration experiment; for commercially available material it is of the order of 0·1. The thickness of the coating must also be known and, if the coating is cast directly on to the part under test, its accurate determination may present practical difficulties.

Assuming that in the oblique incidence observations the angle of incidence is 45° and that Poisson's ratio is the same for the plastic as for the material of the test piece, the separate principal stresses are:

$$\sigma_p = \frac{E_m}{t \,.\, K(1 + \nu)}\left(\frac{\sqrt{2}}{2} r_{ob} - \frac{r_n}{2}\right) \tag{8.6}$$

$$\sigma_q = \frac{E_m}{t \,.\, K(1 + \nu)}\left(\frac{\sqrt{2}}{2} r_{ob} - r_n\right) \tag{8.7}$$

The subscripts *ob* and *n* refer to the retardations in the oblique and normal incidence observations respectively. These equations are given in terms of path retardations and in applying them the following relations will be remembered:

$$r = n \,.\, \lambda \qquad c = \frac{\lambda}{f}$$

where:

r is the path retardation;
n is the fringe order;
λ is the wavelength of light used;
c is the stress-optical coefficient;
f the material fringe value.

Suitable materials and apparatus for the photo-elastic coating method have become available commercially under the name of *Photostress*. For plane surfaces sheet material can be used and for investigations on curved surfaces the material may either be brushed on or cast flat and moulded to the shape of the surface before polymerisation has been completed.

Calibration of the plastic material may be effected by means of a beam specimen to the top of which a small piece of the plastic coating has been bonded. Under the known support and loading conditions, the strain in the plastic can be calculated and the factor K determined from the equation:

$$K = \frac{r}{2t(e_1 - e_2)}$$

e_1 and e_2 being the principal strains in the middle surface of the material. In such an experiment it will usually be necessary to make allowance for the reinforcing effect of the coating and the distance of the middle surface from the surface of the test beam; this is quite easily done by calculating the strains in the composite section by the methods of elementary mechanics. It is claimed that principal strains can be measured to $\pm 10 \times 10^{-6}$ with suitable equipment and calibration techniques.

The coating method of photo-elasticity possesses a number of important advantages: it permits the determination of strains and stresses on the surface of actual structural elements as opposed to transparent plastic models, if desired under service conditions. Readings can be taken over surfaces of almost any degree of complexity and of practically any constructional material. Materials are available which have a linear strain retardation characteristic up to 50 per cent strain, thus permitting measurement within the plastic range. Strain records can be obtained under dynamic loading conditions using ciné-photography or stroboscopic techniques. Long-term stability of the coating material is claimed and the technique is suitable for use under field conditions.

The main limitations of the method are that it can only be used in positions accessible to a light beam (although in this connection it may be noted that observations can be taken from a considerable distance by telescope). Care is necessary to avoid errors resulting from alteration of the calibration constant of the material with temperature and due to differences in the coefficient of expansion of the test piece and coating. If strains are being measured in thin sections it is necessary to make allowance for the reinforcing effect of the plastic coating. It is also possible to use the photo-elastic coating method to measure strains at isolated points. This is done by cementing narrow strips of plastic to the part under investigation at points where it is desired to measure the strain. The strain in the test piece can then be deduced from the fringe order (or colour) developed in the strip. Increased sensitivity can be achieved by putting a stress-raiser, such as a hole, in the strip; this will give a higher fringe order than the plain strip and the correlation between this and the surface strain in the material of the test piece is easily established by calibration experiments.

BRITTLE LACQUER COATINGS

This method of analysis is not based on photo-elasticity, but it is opportune to describe it at this point as it has certain similarities to the photo-elastic coating technique in that it permits the measuring of the magnitude and directions of principal strains on the surface of three dimensional objects and, like the photo-elastic coating method, it is non-destructive.

The idea of examining surface strains by observing the cracking or flaking of a brittle coating on a test object must go back to the earliest days of materials testing. Every student is familiar with the appearance of Luder's lines revealed in a mild steel tensile test piece by the cracking of mill scale on its surface. Valuable information as to the manner of deformation of the test piece is obtained by this observation and over the years attempts have been made to apply this principle in a more general way by applying a brittle coating to the surface of the test object. It is possible to reveal plastic deformations in steel by means of a coating of whitewash. This will crack and flake off when the strain in underlying material reaches something of the order of 2–3 per cent. At a later stage shellac was used, applied either by dissolving it in alcohol or by melting it down on the surface with a gas flame. This is again fairly effective in showing areas of considerable strain, but cannot be relied upon to crack within the elastic range of say steel or aluminium. Materials are now available commercially which will do this and can therefore be used for non-destructive tests on structural elements or machine parts.

A range of brittle coating materials is marketed by the Magnaflux Corporation of America under the name *Stresscoat*. These contain wood resin and dibutyl phithalate with carbon disulphide as a solvent. These and other constituents are mixed in varying proportions to produce lacquers which will crack at repeatable strain values when used under a range of ambient conditions.

Ideally, brittle lacquer experiments should be carried out in a constant temperature and humidity room. Ambient temperature control to within $\pm 2°F$ (1 °C) is necessary for quantitative work and the difference between the temperature of the test piece and the calibration bar must not exceed $\pm 0{\cdot}5°F$ (0·25 °C). If these limits are greatly exceeded the results obtained will be qualitative only.

The lacquer is applied by spray to the surface of the test piece and kept at constant temperature for 12–24 hr before testing. The thickness of the coating should be between 0·1 mm and 0·2 mm. At the same time as the test piece is sprayed, a coating is applied to a small cantilever calibration bar which, at the time of the test, is deformed by depressing the free end with a cam and noting where the first crack appears. The strain at this part can be calculated if the maximum deflection of the bar is known. If the lacquer has been correctly selected to suit the ambient temperature and humidity and assuming that the coating has been correctly applied, the first cracks should appear at a strain of $700–800 \times 10^{-6}$. By observing the extent of cracking at various load increments it will be possible to plot lines along which the tensile strains have about the same magnitude. The cracks may be made more clearly visible for photography by painting a red dye over the surface; this penetrates into the cracks and remains there after the surplus is wiped off.

Lines of compressive strain can be traced by applying the coating when the test piece is under load and then releasing the load after the lacquer has dried out. The two sets of crack lines when fully developed reveal the pattern of the lines of principal stress in the element under test.

The advantages of this method of analysis are: (*a*) that it is a non-destructive test and (*b*) it does not require the use of special models. The disadvantages are: (*a*) that the technique requires a good deal of skill in applying the coating and (*b*) it is relatively insensitive and (*c*) for reliable results very close control over ambient conditions is required. The method has been quite widely applied in relation to machine design where it has been used to obtain information concerning surface strains on complicated three dimensional elements. Like the photo-elastic coating technique, it can be used as a complement to strain gauge analysis. If the position of maximum stress and directions of the principal stresses are determined by the brittle lacquer (or photo-elastic coating) method, the magnitude of the principal stresses can be checked at important points using only a small number of strain gauges.

BIBLIOGRAPHY

Frozen Stress Photo-elasticity

1. W. A. P. Fisher, Basic physical properties relied upon in the frozen stress technique, *Proc. Inst. Mech. Eng.*, **158**, 230–55 (1948).
2. M. Ballet and G. Mallet, On the use of ethoxylene resin in three dimensional photoelasticity for the freezing technique, *C.R. Acad. Sci., Paris*, **233** (16), 846–7 (1951).
3. K. Kuske, Multiphase theory of plastics, *Exp. Mech.*, **2** (9), 278–80 (1962).
4. R. B. Heywood, Modern applications of photoelasticity, *Proc. I. Mech. Eng.*, **158**, 235–40 (1948).
5. M. Hetényi, The application of hardening resins in three dimensional photoelastic studies, *J. App. Phys.*, **10**, 295–300 (1939).
6. M. M. Frocht, Studies in three dimensional photoelasticity, *Trans. A.S.M.E., J. App. Mech.*, **66**, A10–6 (1944). See also *Proc. Soc. Exp. Stress Anal.*, **2** (1), 128–38 (1944).
7. M. M. Frocht, Studies in three dimensional photoelasticity, torsional stresses by oblique incidence, *Trans. A.S.M.E., J. App. Mech.*, **66**, A229–34 (1944).
8. M. M. Frocht *et al.*, Photoelasticity—a precision instrument of stress analysis, *Proc. Soc. Exp. Stress Anal.*, **11** (1), 105–12 (1953).
9. R. C. O'Rourke, Three dimensional photoelasticity, *J. App. Phys.*, **22** (7), 872–8 (1951).

10. M. M. FROCHT and R. GUERNSEY, *A special investigation to develop a general method for three dimensional photoelastic stress analysis*, NACA, Rep. 1148 (1953).
11. M. M. LEVEN, Quantitative three dimensional photoelasticity, *Proc. Soc. Exp. Stress Anal.*, **12** (2), 157–172 (1955).
12. KH. ABEN, On a full determination of the volumetric stressed state by utilising the method of photoelasticity, *Izv. Akad. Nauk. Est. SSR. Ser. Fiz. Mat. i Tech. Nauk*, **9** (2), 134–144 (1960).
13. H. T. JESSOP, A tilting stage method for three-dimensional photoelastic investigations, *Brit. J. App. Phys.*, **8** (1), 30–32 (1957).
14. J. S. BROCK, The determination of effective stress and maximum shear stress by means of small cubes taken from photoelastic models, *Proc. Exp. Stress Anal.*, **16** (1), 1–8 (1958).

Photo-elastic Coatings

15. C. R. SMITH and F. ZANDMAN, Photostress plastic for measuring stress distribution around holes, *Proc. Soc. Exp. Stress Anal.*, **17** (1), 23–24 (1959).
16. T. SLOT, Reflection polariscope for photography of photoelastic coatings, *Proc. Soc. Exp. Stress Anal.*, **19** (1), 41–47 (1961).
17. F. ZANDMAN *et al.*, Reinforcing effects of birefringent coatings, *Proc. Soc. Exp. Stress Anal.*, **19** (1), 55–64 (1961).
18. T. C. LEE *et al.*, Thickness effects in birefringent coatings with radial symmetry, *Proc. Soc. Exp. Stress Anal.*, **18** (2), 134–40 (1960).
19. G. V. OPPEL, Photoelastic strain gauges, *Proc. Soc. Exp. Stress Anal.*, **18** (1), 65–73 (1960).
20. J. DUFFY, Effects of the thickness of birefringent coatings, *Proc. Soc. Exp. Stress Anal.*, **18** (1), 74–82.
21. J. SCHWARZHOFER, Applications of photoelastic strain gauges, *Exp. Stress Anal.*, **18** (2), 198–202 (1960).
22. H. FESSLER and D. J. HAINES, A photoelastic technique for strain measurement in flat aluminium alloy surfaces, *Brit. J. App. Phys.*, 282–87 (1958).

Brittle Lacquer Coatings

23. W. F. STOKEY, Elastic and creep properties of Stresscoat, *Proc. Soc. Exp. Stress Anal.*, **10** (1), 179–87 (1952).
24. F. N. SINGDALE, Improved brittle coatings for use under widely varying temperature conditions, *Proc. Soc. Exp. Stress Anal.*, **11** (2), 173–78 (1953).
25. A. J. DURELLI *et al.*, Study of some properties of Stresscoat, *Exp. Stress Anal.*, **12** (2), 55–76 (1954).
26. J. H. CUNNINGHAM and J. M. YAVARSKY, The brittle lacquer technique of stress analysis as applied to anisotropic materials, *Proc. Soc. Exp. Stress Anal.*, **16** (2), 101–8.

27. G. W. Brown and D. M. Cunningham, Calibration of Stresscoat by impact, *Exp. Stress Anal.*, **19** (1), 150–54 (1961).

IX

STRUCTURAL MODEL ANALYSIS

A NUMBER of experimental techniques have been evolved for the determination of bending moments, shearing forces and deflections in elastic structures by the use of comparatively simple models. These methods of model analysis can be grouped under two headings, namely, direct and indirect, depending upon whether the loading on the actual structure is reproduced on the model and measurements made on it to evaluate the resulting strains, etc., or whether the model is used to find influence coefficients, lines or surfaces, from which the moments, etc., due to the actual loading can be determined. The methods used for model analysis are summarised in Table 9.1.

It will be appreciated that there is a distinction between the detailed exploration of stress in a structural detail and the determination of total moments and forces acting in a member. The latter is in view in the present chapter, although it will be observed that in some cases the same experimental techniques may be used for both types of investigation.

DIRECT METHODS

Analysis by deflection measurement

Deflection measurements, *per se*, are not particularly useful as a means of determining stresses in structures, except in a few special cases. For example in a beam, an infinite number of bending moment curves can yield practically the same deflection curve. In other words, it is seldom safe to differentiate a deflection curve twice so as to determine the bending moment

Table 9.1

A. Direct Methods	
(1) Methods based on slope or deflection measurements:	Direct measurement of slopes and deflections Moiré fringe method Moment indicator
(2) Strain measurement:	Measurement of strains by resistance strain gauges
(3) Photo-elasticity :	
B. Indirect Methods	
Based on Maxwell–Betti and Müller–Breslau theorems	

curve. The only exception to this occurs when it is known that the deflection curve is the same shape as the bending moment diagram, in which case it is competent to deduce stresses from deflection. If, however, readings are taken so as to measure the rotations of the joints in a framed structure, it is possible to deduce the bending moments and shearing forces in its members. One possible method[1] of doing this is shown in Fig. 9.1; deflections are measured with micrometer microscopes on light target attached to the model by stiff wires. The rotations are obtained from these, knowing the distance of the target from the joint concerned. The moments are then obtained from the slope deflection equation:

$$M_{AB} = \frac{2EI}{L}\left(2\theta_A + \theta_B - \frac{3\delta}{L}\right) \tag{9.1}$$

$$M_{BA} = \frac{2EI}{L}\left(2\theta_B + \theta_A - \frac{3\delta}{L}\right)$$

In an experiment of this kind the model material may be metal or plastic and the displacement can be measured with a micrometer eye-piece microscope. Loads are applied by dead weight, using strings and pulleys if necessary. It is sometimes an advantage to mount the model horizontally, in which case it should be supported on ball bearings in order to eliminate errors due to friction.

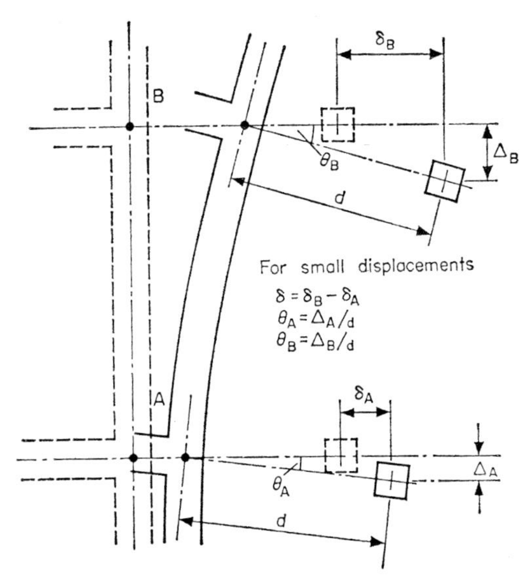

Fig. 9.1. Determination of moments from displacement measurements.

If the model is of plastic, it is necessary to guard against errors arising from mechanical creep in the material. Possible ways of dealing with this problem have been discussed in Chapter I (p. 6). A further possibility in structural model experiments is to apply the load through a compensating balance which consists of a U-shaped member, as indicated in Fig. 9.2(a), cut from the same sheet of material as the model, and thus possessing identical creep properties. Referring to Fig. 9.2(b), the force P is the same on the balance and on the model and the displacements of the model and of the balance at any specified time, t, are δ_m and δ_b respectively. Then:

$$\delta_m = K_m \cdot \frac{P_t}{E_t} \qquad \delta_b = K_b \cdot \frac{P_t}{E_t} \tag{9.2}$$

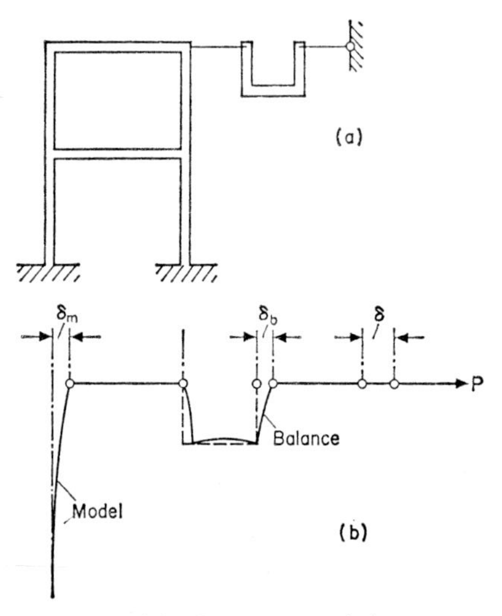

Fig. 9.2. Compensating balance.

where K_m and K_b are constants depending only on the geometry of the model and of the balance and E_t is the effective modulus of elasticity at time t. We note that:

$$\delta = \delta_m + \delta_b = (K_m + K_b)\,\frac{P_t}{E_t} \tag{9.3}$$

or

$$\frac{\delta}{K_m + K_b} = \frac{P_t}{E_t} \tag{9.4}$$

This means that the ratio of P_t/E_t does not vary with time for a given value of the total displacement δ, that is P_t and E_t change proportionally. Furthermore, substituting in (9.2)

$$\delta_m = \frac{K_m}{K_m + K_b} . \delta \quad \text{and} \quad \delta_b = \frac{K_b}{K_m + K_b} . \delta \tag{9.5}$$

which shows that δ_m and δ_b remain constant with time. In an experiment in which deflections only are measured this means that the use of the compensating balance effectively eliminates creep errors. If strains are being measured in the model, stresses can be found from there as follows: if the strain at some point in the model corresponding to δ is e_m then:

$$\sigma_m = E_t . e_m \tag{9.6}$$

From (9.2):

$$E_t = \frac{K_b}{\delta_b} . P_t \tag{9.7}$$

so that:

$$\sigma_m = \frac{K_b}{\delta_b} . P_t . e_m \tag{9.8}$$

Thus, the stresses can be determined without reference to Young's modulus. The constant K_b can be calculated or found experimentally. The deflection δ_b is measured and the stresses can then be evaluated for any convenient value of P_t.

Instead of measuring displacements with a microscope the moment indicator described in Chapter II (p. 23) may be used. The compensating balance can, of course, be used in conjunction with the moment indicator.

The Moiré Method

Another direct method of analysis based on deflection and slope measurements is the moiré fringe technique for the determination of the stresses in plates [3, 4, 10]. At a point in a transversally loaded elastic plate, the bending moments per unit length M_x, M_y, and M_{xy}, are related to the deflection w, by the following differential equations:

$$M_x = -D\left(\frac{\partial^2 w}{\partial x^2} + \nu . \frac{\partial^2 w}{\partial y^2}\right) \tag{9.9}$$

$$M_y = -D\left(\frac{\partial^2 w}{\partial y^2} + \nu \, . \, \frac{\partial^2 w}{\partial x^2}\right) \tag{9.10}$$

$$M_{xy} = m_{yx} = -D(1 - \nu)\frac{\partial^2 w}{\partial x \, . \, \partial y}$$

where D is the flexural rigidity per unit length, i.e.

$$D = \frac{Eh^3}{12} \Big/ (1 - \nu^2)$$

h being the thickness of the plate.

The first derivative of the deflection surface is of course its slope in the direction of the particular axis of reference and it is the slopes that are measured in the moiré fringe method, the bending moments being determined by differentiating the slope curves on selected sections. The slope measurement is effected by a reflection technique and by a rather ingenious procedure it is possible to obtain contours of $\partial w/\partial x$ and $\partial w/\partial y$ for the whole of a loaded model. The method will be understood by referring first of all to Fig. 9.3, which is a

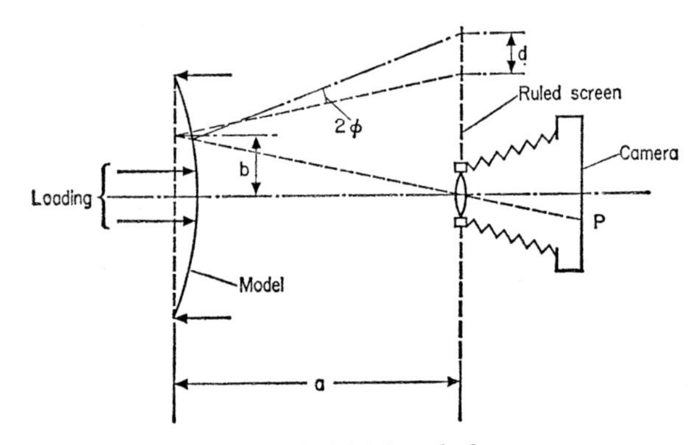

Fig. 9.3. Moiré method.

cross section of the experimental set-up. As shown here, the model is mounted vertically in a loading frame. Parallel to the plane of the model and at a distance from it is a ruled screen at the centre of which a camera faces the model. The surface of the model nearest the camera is reflective so that an image of the ruled screen appears on the camera focusing screen or photographic plate. Suppose that an image of a particular line on the ruled screen appears at P on the photographic plate before load is applied to the model and that after loading, an adjoining line distance d from the first is observed at P. Then the slope of the model can be found from the relationship:

$$\phi = \frac{d}{2a[1 + (b/a)^2]} \tag{9.12}$$

If instead of a flat ruled screen a suitably curved one is used, it is possible to eliminate the term $(b/a)^2$ in (9.12) and the value of ϕ is then simply $d/2a$. A cylindrical screen whose radius is $3{\cdot}75a$ is sufficiently close for this purpose; the lines on the screen are, of course, ruled parallel to the axis of the cylinder.

Further inspection of Fig. 9.3 will show that as the curvature of the plate is increased, a ray from a particular point on the camera plate will traverse the ruled screen "seeing" alternately, lines and spaces. Thus considering a small element of the model, the initial image will be as in Fig. 9.4(a); the reflected pattern will be displaced as the deflection is increased until at a certain stage, the image will have been displaced relative to the initial image by an amount equal to the width of one of the lines. If photographs are taken of the initial and final appearance of the ruled screen they will be as shown in Figs. 9.4(a) and (b); and if they are superimposed the lines in (b) will fall into the spaces in (a); the result will have a uniformly dark appearance as in (c). Superposition of negatives taken at other deflections will show partial darkening of the field. If a strip of the model plate is now examined, the superimposed negatives will show alternate light and dark areas corresponding to places where the images coincide and interfere. If the curvature of the plate is uniform, i.e. cylindrical, the moiré fringes so formed will be uniformly spaced but in general, the spacing will vary across the width of the plate.

Clearly, a light zone will appear whenever the image is displaced by a distance equal to the spacing of the lines. The spacing of the centre of the light and dark areas in turn corresponds to a displacement equal to the spacing of the lines on the screen.

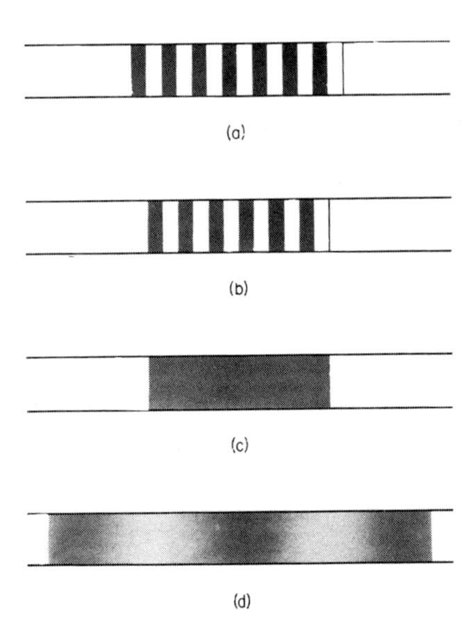

Fig. 9.4. The development of moiré fringes.

Extending this discussion to a uniformly loaded rectangular plate supported round its edges, it is not difficult to see that the initially straight, parallel lines (Fig. 9.5(a)) will be distorted into a cushion-like pattern (Fig. 9.5(b)) when the plate is loaded. Superposition of the two patterns yields the moiré fringe pattern of Fig. 9.5(c). These fringes are contours of equal displacement of the lines in the direction at right angles to the ruling. If the ruled screen is curved, as described above, the moiré fringes are also contours of equal slope in this direction.

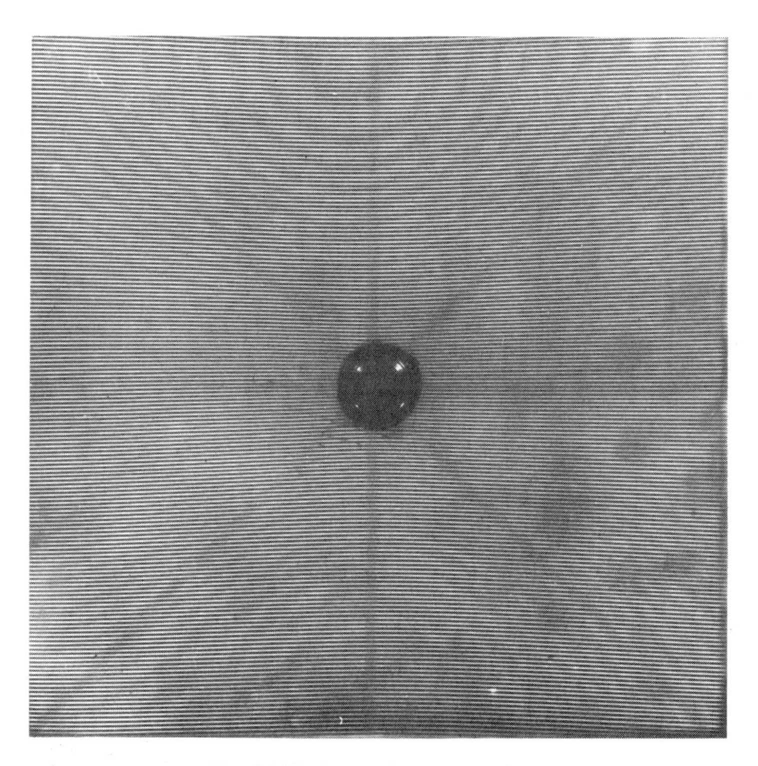

Fig. 9.5(a). Moiré fringes in a plate.

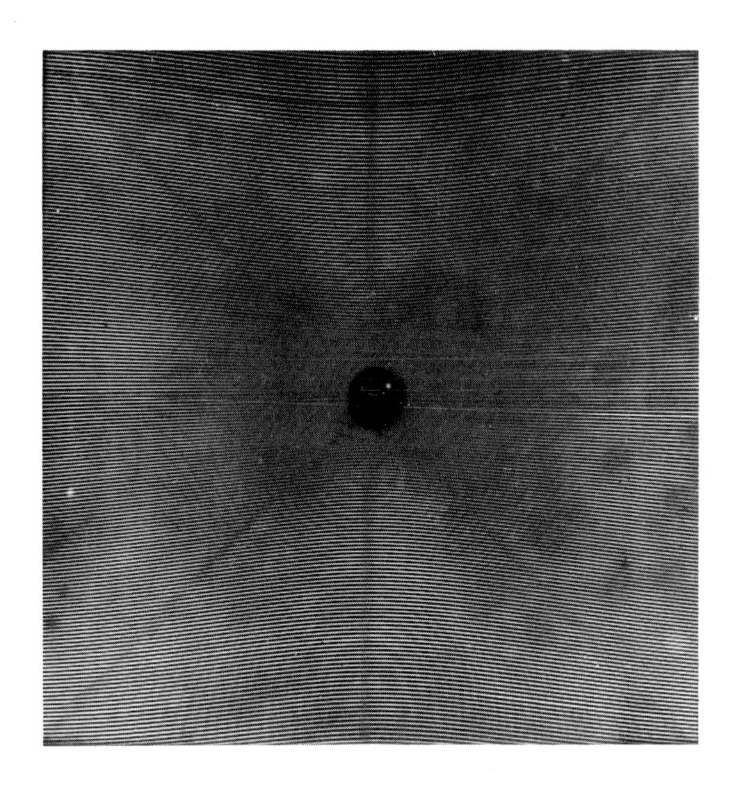

Fig. 9.5(b). Moiré fringes in a plate.

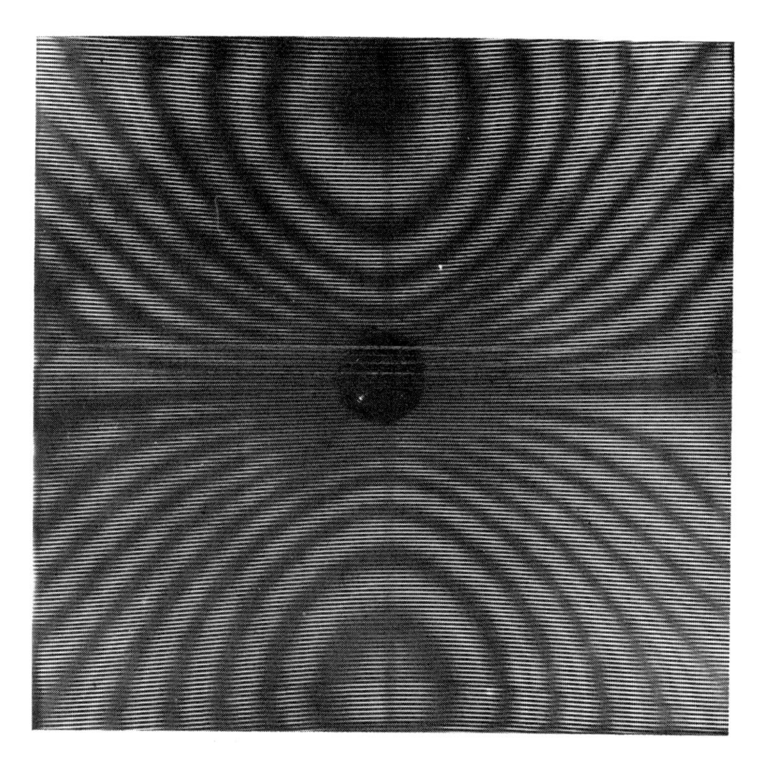

Fig. 9.5(c). Moiré fringes in a plate.

Each fringe will correspond to a particular displacement but it will not be known from the photograph which fringe corresponds to zero or to some multiple of $d/2a$ where d is the spacing of the lines and a the distance of the screen from the model. This, however, is no obstacle since only the derivative of the slope curve along selected lines is required and this can be found without knowing the absolute values of the ordinates of the curve.

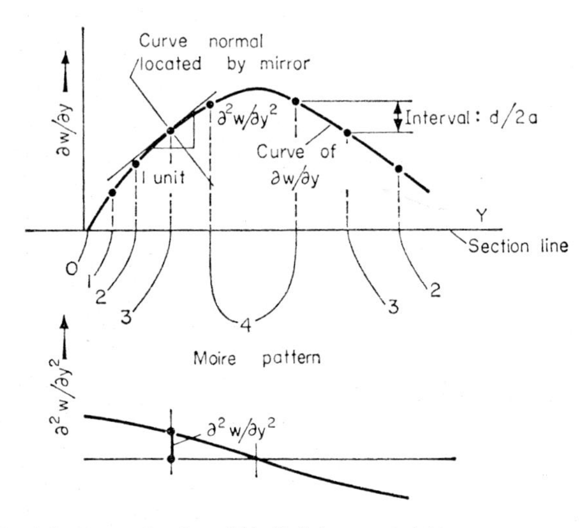

Fig. 9.6. Determination of $\partial^2 w/\partial y^2$ from moiré fringe pattern.

The procedure in the moiré method of analysis will by now be clear: to find the bending moments in a plate model, it is necessary to obtain moiré fringe patterns for two orthogonal settings of the ruled screen, that is with the rulings parallel to the X-axis in one case and to the Y-axis in the other. The first setting gives $\partial w/\partial y$ and the second $\partial w/\partial x$. The fringe patterns are produced by photographing, on the same plate, the ruled screen as reflected by the model before and after application of the load. It may be observed that initial lack of flatness of the plate, leading to distortion of the lines, will not affect the moiré pattern as this appears as the result of the difference between the initial and final states.

From the fringe photographs, curves of $\partial w/\partial y$ and $\partial w/\partial x$ are plotted along selected lines and differentiated in turn to give $\partial^2 w/\partial y^2$, $\partial^2 w/\partial x^2$ and $\partial^2 w/\partial x \,.\, \partial y$.

Any convenient method of graphical differentiation may be employed, the most obvious one being to measure the slope of the tangent at a suitable number of points along the curve. A small rectangular mirror laid across the curve and held at right angles to the paper (as indicated in Fig. 9.6) will permit quite accurate location of the normal to the curve at these points. Having found the derivatives in this way, the moments m_x, m_y and m_{xy} are found from equations (9.9)–(9.11). The value of D can be calculated from the elastic constants of the plate and its thickness or may be determined experimentally from a test on a simple calibration plate for which a theoretical solution exists.

The apparatus for the moiré method is comparatively simple and need not be expensive. The distance of the screen from the model and the ruling on the screen should be chosen so that the ratio $d/2a$ is about 0·002. A satisfactory line width is about 1·5 mm. It is of course essential that the lines should be of uniform thickness and spacing, as errors in this direction would of course upset the accuracy of the results. With reasonable care, an accuracy of ± 5 per cent in moment determinations can be attained which is sufficient for most practical purposes.

The moiré method is perhaps the most convenient experimental technique for investigating plate problems. Strain gauges can be used but rather large numbers are required in this case. Modifications of the photo-elastic method have also been tried but are difficult and tedious. A more satisfactory alternative is the direct measurement of plate curvature; this, however, will be less convenient than the moiré method as a large number of separate observations will be needed for a complete exploration of the moments in a plate.

Fig. 9.7. Moiré method; apparatus.

Strain Gauges

Electrical resistance strain gauges are extensively used for direct model analysis, particularly of dams, bridges, decks and complex building structures. The instrumentation and technique relating to resistance strain gauges has been fully discussed in Chapter IV and examples of their use abound in the literature.

Photo-elasticity

This method is primarily of value for the stress analysis of structural details but it can be used for the analysis of entire structures and indeed in some cases provides the only feasible model technique. The method has been described in Chapter VII and VIII and it is only necessary at this point to indicate by example the kind of problems in which photo-elasticity may be useful.

Certain structural frameworks are very easily analysed by photo-elasticity by locating points of inflection in the members. Fig. 9.8 shows the appearance of a panel of a typical Vierendeel girder in the polariscope. As may be seen, the characteristic appearance of the fringes makes it possible to locate the points of inflection with reasonable accuracy. Once these have been located, the structure can be subdivided into a number of statically determinate parts for the calculation of direct forces, shears and bending moment in the members.

Photo-elasticity has also proved useful in the analysis of dams, using a centrifuge to simulate gravitational loading (cf. Chapter XII, p. 180).

INDIRECT METHODS

In the direct methods of analysis so far described, the actual loading on the structure is represented and measurements are made of the resulting deflections or strains. In the indirect methods, however, no attempt is made to reproduce the actual loading but instead, certain deformations are imposed on the model and from the resulting deflections influence lines, surfaces or coefficients are determined from which the bending moments, shearing forces, etc., produced by the actual loading can be calculated [5].

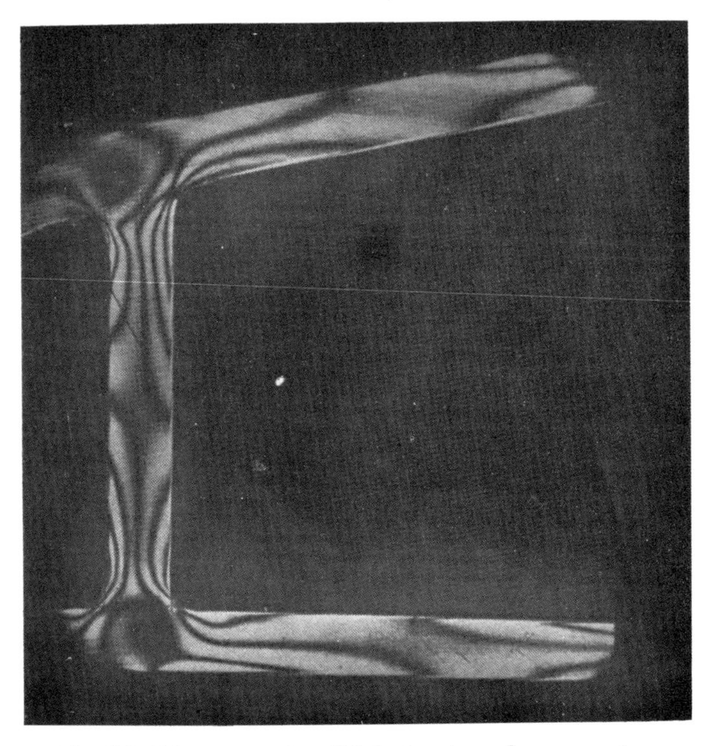

Fig. 9.8. Fringes at point of inflection in a flexure member.

Various techniques have been used, but all depend on the application of the Maxwell–Betti and Müller–Breslau theorems. Referring to Fig. 9.9 the Maxwell–Betti theorem states that, in an elastic structure, the deflection at A along a line AC due to a load at another point B in direction DB, is equal to the deflection at B along BD when the same load is applied at A in direction CA. In short:

$$\delta_{AB} = \delta_{BA}$$

This is the well known reciprocal deflection relation.

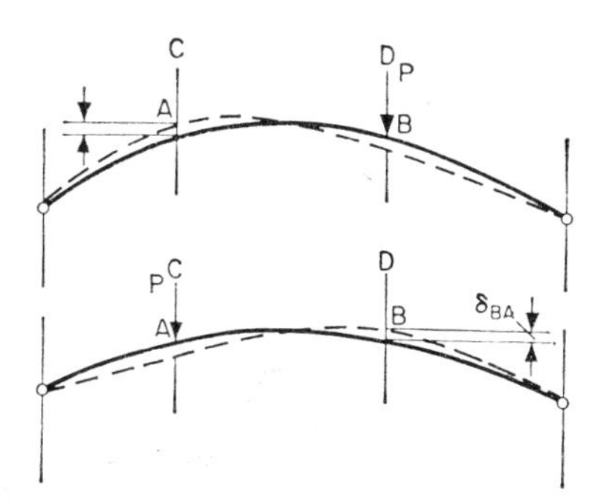

Fig. 9.9. Reciprocal deflection relation.

The Müller–Breslau theorem postulates that in a statically indeterminate frame, the deflection of the centre line of the frame produced by displacing one of the supports in the direction of a redundant reaction defines the influence line for that reaction to some scale. This theorem also applies to cuts made in members within the frame.

The basic procedure in indirect model analysis will be understood by referring to the continuous beam shown in Fig. 9.10. If the influence line for R_B is required, the support at B is removed and point B is displaced by an arbitrary amount δ_{BB} in the direction of R_B. It may be assumed that this displacement corresponds to unit load at B. Then the vertical displacement at any point X is δ_{XB}, i.e. the displacement of X due to unit load at B. From the Maxwell–Betti theorem, $\delta_{XB} = \delta_{BX}$, that is the displacement of B due to unit load at X is equal to the displacement of X due to unit load at B. Now the displacement at B is zero when the support at B is in position and in order to cancel out the displacement δ_{BX}

caused by unit load at X we must apply the reaction R_B. Since δ_{BB} is the displacement at B due to unit load at that point:

$$R_B \,.\, \delta_{BB} = \delta_{BX}$$

so that

$$R_B = \frac{\delta_{BX}}{\delta_{BB}}$$

Thus, having determined the displacements δ_{XB} at various points along the beam, the ordinates of the required influence line is found by dividing these deflections by δ_{BB}. Exactly the same procedure can be followed to find R_A as shown in Fig. 9.10(b).

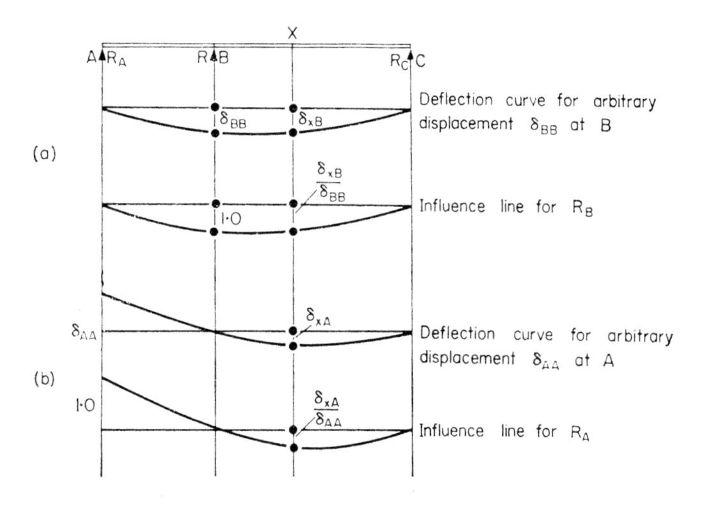

Fig. 9.10. Influence lines for a continuous beam.

The indirect method was first put forward by G. E. Beggs [(6)] who developed apparatus for imposing small displacements on celluloid models by means of plugs and wedges (Beggs deformeter); the resulting displacements are measured with an eyepiece micrometer microscope. Although this is without doubt the most accurate version of the indirect method, it is possible to use models which are sufficiently large and flexible to permit measurements to be made with a steel rule reading to 0·5 mm. The models may be made from steel splines as in Gottschalk's "Continostat" and Rieckhof's "Nupubest"[2] or of celluloid or Perspex as in the methods developed by Pippard and Sparkes[5]. This latter method is more adaptable than those using steel splines because any moment of inertia can be represented. The experiment can be carried out on a sheet of smooth paper on a drawing board, the support reactions being provided by map pins. Fine holes are drilled along the axis of the structure so that the initial and deflected centre lines can be marked on the paper with a needle. To improve the accuracy of measurement and to reduce errors due to change of geometry resulting from the deformation of the structure, it is useful to apply equal deformations about both sides of the undeflected centre line and to measure deflections between the two curves thus obtained; this procedure will be understood from Fig. 9.11 which shows the determination of the influence line for horizontal thrust in a two pinned portal frame for vertical loads on the beam.

The same general procedure is followed if it is required to determine a redundant support moment: in this case an arbitrary rotation is imposed in place of the redundant moment, as indicated in Fig. 9.12. If the influence lines for M_b due to vertical loads on the beam are required, the ordinates are obtained by dividing the vertical deflections such as δ_x by the rotation ϕ_B.

If it is required to find, say, the bending moment or shear at a section within a member, it is necessary to cut the member and impose a deformation, angular or linear as the case may be, on the centre line at the section concerned. Deflections are then measured at various points and the influence line ordinates determined as previously described. In structures which are externally redundant only, it will not be necessary to resort to this procedure as it will be possible in this case to construct

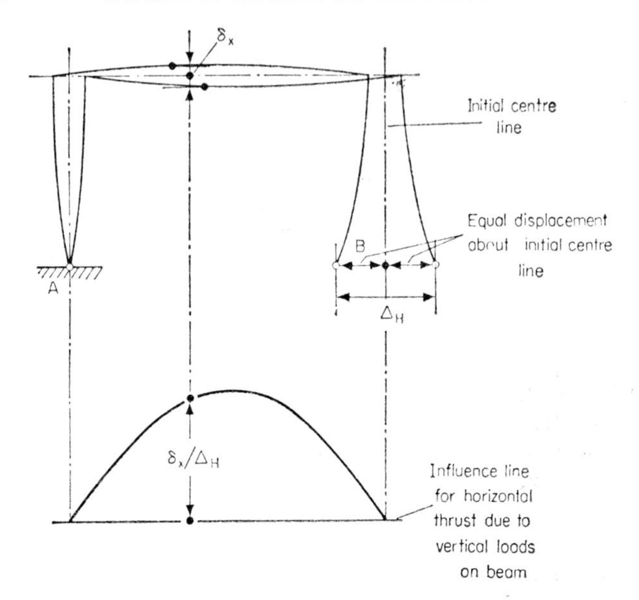

Fig. 9.11. Influence line for horizontal thrust in a portal frame.

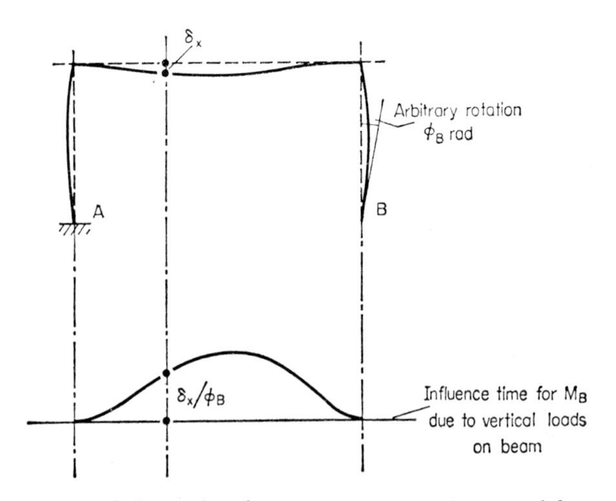

Fig. 9.12. Influence line for support movement in a portal frame.

the influence lines for moments, etc., at internal sections once those for redundant reactions have been determined.

A useful procedure in the indirect method as applied to plane frames has been described by Rocha and Borges[7]. This consists of photographing the model before and after deformation on the same photographic plate. The displacements can then be measured from the photographic record at leisure; as well as giving improved accuracy as compared with direct measurement on the model, this method provides a permanent record of the test.

The indirect method can be applied to more complicated structures and has been effectively used, for example, in the analysis of braced domes[8]. In such a case, the procedure is to make a model in which the length of the members (or a sufficient number of them) can be altered by a known amount. If the effect of vertical loading is being considered, the vertical deflections of the node points due to an arbitrary shortening of one of the members are measured; these deflections divided by the imposed alteration in the length of the member give the node point influence coefficients for load in the member whose length was altered. The method has also been successfully applied to the analysis of three dimensional, reinforced concrete slab structures[9].

BIBLIOGRAPHY

1. A. J. S. Pippard, *The Experimental Study of Structures*, p. 29. Edward Arnold, 1947.
2. M. Hetényi (Ed.), *Handbook of Experimental Stress Analysis*, pp. 657–69. Wiley, 1950.
3. F. K. Ligtenberg, The Moiré method—A new experimental method for the determination of moments in small slab models, *Proc. Soc. Exp. Stress Anal.*, **12** (2), 83–98 (1955).
4. C. G. J. Vreedenburgh, New progress in our knowledge about the moment distribution in flat slabs by means of the Moiré method, *Proc. Soc. Exp. Stress Anal.*, **12** (2), 99–114 (1955).
5. A. J. S. Pippard and S. R. Sparkes, Simple experimental solution of certain structural design problems, *J. Inst. C.E.*, **4**, 79–92 (1936–37).

6. G. E. Beggs, An accurate mechanical solution of statically indeterminate structures by use of paper models and special gauges, *Proc. American Concrete Inst.*, **18**, 58–78 (1922).
7. A. Rocha and J. F. Borges, Photographic method for model analysis of structures, *Proc. Soc. Exp. Stress Anal.*, **8** (2), 129–142 (1951).
8. Z. Makowski and A. J. S. Pippard, Experimental analysis of space structures, with particular reference to braced domes; with a note on stresses in supporting ring-girders, *Proc. Inst. C.E.*, **1** (III), 420–441 (1952).
9. K. H. Gerstle and R. W. Clough, Model analysis of three dimensional slab structures, *Proc. Soc. Exp. Stress Anal.*, **13** (2), 133–140 (1956).
10. A. J. Durelli and I. M. Daniel, Structural model analysis by means of moiré fringes, *Proc. ASCE*, **86**, ST12 (J. Struct. Div.), 93–102 (1960).

X

SPECIAL INSTRUMENTS FOR DYNAMIC STRESS ANALYSIS

REFERENCE has been made in Chapters IV and V to the measurement of dynamic strains by various forms of electrical gauges and in Chapter II to the application of cyclic loads to structures by mechanical or electrical means. In addition to this equipment, investigations of dynamic behaviour of structures will generally require instruments capable of detecting and recording accelerations and deflections under dynamic loading. Deflection measuring instruments may be of the directly connected type, attached to a fixed base as in static work, or of the seismic type, attached to the moving part and detecting its movement relative to a suspended mass which remains stationary by reason of its inertia. Accelerations can also be measured by seismic instruments.

DIRECTLY CONNECTED DEFLECTION INSTRUMENTS

The simplest direct reading deflection instrument is the ordinary dial gauge. Although intended for use under static conditions, a dial gauge can, within limits, be used to measure the amplitude of a steady vibration. When used for this purpose, the deflection reading can be obtained by observing the arc of movement of the dial gauge pointer over the scale; generally this will appear as a blur, the limits of which can be quite accurately read. The measurement is only valid if the plunger of the dial gauge remains in contact with the test object and if there are no errors due to backlash in the mechanism. Backlash is normally taken up by a hair spring and errors from this source appear when the inertia forces in the moving parts of the gauge overcome the torque exerted by the hair spring. For a given dial gauge this will take place at a definite frequency and amplitude. Weiss [1] has found that the readings are accurate until:

$$\left(\frac{\omega}{p}\right)^2 = 1 + \frac{2(\Delta + \delta)}{x} \tag{10.1}$$

where p is the natural circular frequency of the dial gauge mechanism;

ω is the circular frequency of the simple harmonic motion being measured;

x is the amplitude of the s.h.m. at audible loss of contact;

Δ is the initial total compression of the gauge spring (in inches or cms);

δ is the constant friction force of dial gauge mechanism/spring constant of dial gauge.

The natural frequency of the gauge and the quantity $(\Delta + \delta)$ can be found experimentally by determining for any two frequencies the amplitudes at which there is audible separation of the dial gauge plunger from the table of a calibrating device. If these are respectively x_1, ω_1, and x_2, ω_2, then:

$$p^2 = \frac{x_1\omega_1^2 - x_2\omega_2^2}{x_1 - x_2} \tag{10.2}$$

and

$$2(\Delta + \delta) = \frac{x_1 x_2(\omega_2^2 - \omega_1^2)}{x_1\omega_1^2 - x_2\omega_2} \tag{10.3}$$

It is assumed that Δ is the same for both observations and that $x_1 > x_2$ and $\omega_1 < \omega_2$.

Dial gauges are limited to the measurement of the amplitude of steady simple harmonic motions within their frequency range and obviously do not provide a record of the motion. A number of mechanical recording instruments are available such as the Cambridge Universal Vibrograph (see p. 140). Although this is primarily a seismic instrument it can be used to record directly transmitted movements on a celluloid tape. The amplitude of the movement is reduced by a factor of 10 or 50 according to the stylus fitted but the trace is magnified for measurement in a simple projection device supplied with the instrument. A timing signal is recorded alongside the deflection trace and an electrically operated event marker is also fitted. This device can be used to record transient effects as well as steady vibrations.

Electrical instruments are inherently very suitable for dynamic deflection measurements up to very high frequencies. They can be made so as to impose negligible restraint on the movement of very light mechanical systems and recording is comparatively easy using readily available instruments. Inductance and capacitance gauges, described in Chapter V, are easily adapted to serve as pickups and can be made to indicate on a cathode ray oscilloscope from which a photographic record can be obtained. Inductance gauges are also suitable for connection to a recording galvanometer; this being an electromagnetic device having a relatively low input impedance, it requires appreciable power from the pickup.

PRINCIPLE OF SEISMIC VIBRATION INSTRUMENTS

Seismic instruments are constructed on the same principle as the seismographs used to record earthquakes. They have a suspended mass of high inertia constrained to move in a particular direction; the motion of the case or base of the instrument relative to the suspended mass is indicated or recorded. Depending upon the characteristics of the system, this relative motion is a reproduction of the displacement or of the acceleration of the body to which the case is secured. It is also possible to devise a seismic instrument to measure the velocity of motion.

Referring to Fig. 10.1, the differential equation describing

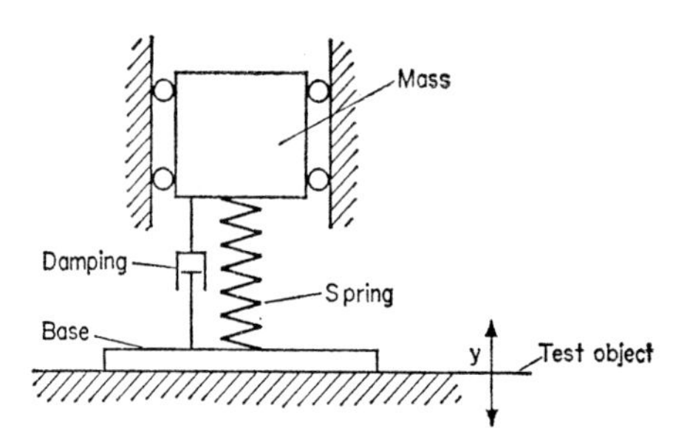

Fig. 10.1. Sprung mass system.

the relative displacement of the suspended mass and the base of a seismic pickup is:

$$\frac{d^2x}{dt^2} + 2np\frac{dx}{dt} + p^2x = -\frac{d^2y}{dt^2} \tag{10.4}$$

where y is the displacement of the test object and base;
x is the displacement of the mass relative to the base;
m is the mass of seismic weight;
k is the spring constant;
c is the equivalent viscous damping constant;
$C_c = 2\sqrt{(k \,.\, m)}$ is the critical damping constant;
$n = C/C_c$: damping ratio;
$p = \sqrt{(k/m)}$: natural circular frequency of the spring–mass system.

If the test object on which the instrument is mounted is performing a simple harmonic motion:

$$y = y_0 \sin \omega t \tag{10.5}$$

the solution of (10.4) is:

$$x = y_0 \frac{(\omega/p)^2}{\sqrt{\left[4n^2\left(\frac{\omega}{p}\right)^2 + \left(1 - \left(\frac{\omega}{p}\right)^2\right)^2\right]}} \cdot \sin\left(\omega t - \tan^{-1} \frac{2n\,\frac{\omega}{p}}{1 - (\omega/p)^2}\right)$$

If $\omega/p > 4$ and $n < 0{\cdot}1$ then:

$$x = -y_0 \sin \omega t \tag{10.6}$$

That is, the relative displacement of the mass and base is very nearly the same as the motion of the test piece.

If $\omega/p \leq 0{\cdot}6$ and $n \simeq 0{\cdot}6$:

$$x \simeq \frac{1}{p^2} \cdot y_0 \omega^2 \sin \omega t \tag{10.7}$$

In this case the relative displacement is seen to be proportional to the acceleration of the test piece.

It will be seen from this that any seismic pickup will function either as a deflection measuring device or as an accelerometer according to the frequency range in which it is used. In practice, deflection measuring instruments (referred to as vibrographs, pallographs or seismographs) are built with a very "soft" suspension to make p small and with as little damping as possible. Accelerometers, on the other hand, are built with stiff springs to give a high value of p and with some form of damping arranged to make n as nearly 0·65 as practicable. In the range $\omega/p \simeq 0{\cdot}7$ to $\omega/p \simeq 4$, that is, when the frequency of the disturbing vibration is in the region of the natural frequency of the sprung mass, the relative displacement of the mass and base will reflect neither the displacement nor the acceleration of the test piece, and if used in this range the records would have to be corrected for dynamic magnification and phase shift. Clearly, it will be more feasible in a given case to select an instrument whose natural period does not come within the resonance range.

In so far as equations (10.6) and (10.7) are approximations, the records obtained from seismic vibration measuring instruments will be to some extent distorted and may require the application of corrections[3, 4, 5] to remove these errors. In well designed and carefully calibrated instruments used in appropriate frequency ranges, distortion of the trace should not be serious.

Various methods are used in commercial instruments for measuring and recording the relative displacement between the case and suspended mass in a seismic instrument. The simplest system is a lever arrangement which moves a scriber over the surface of a celluloid strip as in the Cambridge Vibrograph, the use of which as a directly connected instrument has already been mentioned. In this instrument, shown in Fig. 10.2, the seismic mass is suspended on two strips of spring steel so as to be sensitive either to horizontal or vertical vibrations according to the direction of mounting. The motion of the mass relative to the frame of the instrument is picked up by a stylus bar which scribes a trace on a celluloid ribbon. The celluloid is drawn past the scriber at a constant rate by a clockwork mechanism. A timing device marks 1/10 sec intervals on the record alongside the deflection trace and another scriber can be operated electrically to indicate external events. The frequency range of this vibrograph is from 10 Hz to 100 Hz, but the low frequency limit can be extended by the use of a special attachment to 1 Hz.

Electrical devices can be also used to follow the relative motion of the case and the seismic mass and are commonly employed in accelerometers. In a typical instrument the sprung mass is mounted so as to form an armature between two coils as shown in Fig. 10.3. These coils are energised with alternating current of considerably higher frequency than the signal to be measured and are connected in a suitable bridge circuit to an oscilloscope or recording galvanometer.

Fig. 10.2. Cambridge Vibrograph.

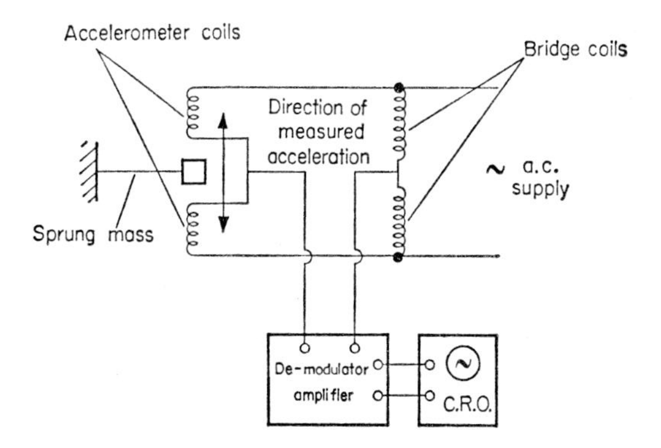

Fig. 10.3. Electrical inductance accelerometer.

Although less frequently required, velocity measurements can be obtained by means of a seismic instrument in which the suspended mass is a magnet; a coil is attached to the case close to the magnet so that the motion of the magnet relative to the coil induces a voltage in the latter which is proportional to their relative velocity. The induced voltage can be recorded on a high impedance oscilloscope.

Vibration instruments are calibrated by attaching them to a table which can be subjected to a simple harmonic motion of known amplitude and frequency. This motion can be induced by either a mechanical or an electrical drive. In mechanically driven calibrators, the rotary motion of the driving motor is converted to a linear oscillation by means of cams, eccentrics, connecting rods or other suitable mechanisms. Inertia forces in such a device are considerable and the machine has to be firmly anchored to a heavy foundation. An electrically driven calibrator for small instruments is shown in Fig. 10.4.

Here, the instrument under test is attached to a table carried by two three-wire suspensions. The drive is provided by a 15 watt loudspeaker unit, the whole assembly being mounted on a heavy concrete base. A signal generator supplies the loudspeaker unit with alternating current of known frequency through a suitable amplifier. The amplitude of the motion is measured by means of a micrometer microscope, sighted on an illuminated target attached to the accelerometer. The maximum acceleration to which the pick-up is subjected is easily calculated from the expression:

$$\frac{d^2x}{dt^2} = x_0\omega^2 \sin \omega t \qquad (10.8)$$

the amplitude x_0 and circular frequency ω, of the motion being known.

CALCULATION OF VIBRATIONAL STRESSES

It is not possible to determine stresses directly from vibration amplitudes with any great accuracy as this would require double differentiation of the test record. Stresses can, however, be determined from acceleration measurement with better accuracy, as may be seen by considering the case of a vibrating beam in which:

$$\frac{d^2M_x}{dx^2} = \frac{\omega_x \cdot a_x}{g}$$

where M_x is the maximum bending moment at x from the end of the beam;
ω_x is the intensity of loading at x;
a_x is the maximum acceleration at x;
g is the gravitational acceleration.

Then the stress at x is:

$$\sigma_x = \frac{M_x \cdot Z_x}{I_x} = \frac{Z_x}{I_x}\left[\iint_0^x \frac{\omega_x a_x}{I_x}\, dx dx + S_0 x + M_0\right]$$

where Z_x is the extreme fibre distance of cross-section at x;
I_x is the moment of inertia of section at x;
S_0 = shear at $x = 0$;
M_0 = bending moment at $x = 0$.

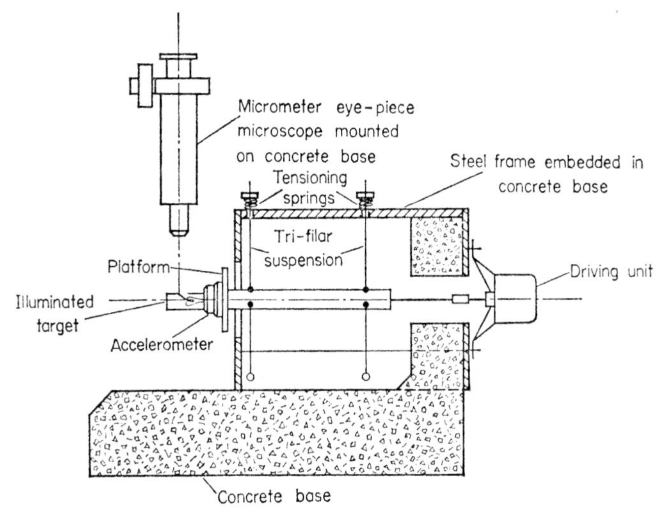

Fig. 10.4. Calibrator for accelerometers.

BIBLIOGRAPHY

1. H. K. WEISS, Errors of the dial gauge as an instrument for measuring amplitudes of vibration, *Rev. Sci. Instrum.*, **9**, 365–369 (1938).
2. H. C. ROBERTS, *Mechanical Measurements by Electrical Methods*. Instruments Publishing Co., Pittsburgh, Pa., 1946.
3. K. KLOTTER, *Messung mechanischer Schingungen* (*Dynamic die Schwingungsmessgeräte*). Springer Verlag, Berlin, 1943.
4. C. S. DRAPER and O. P. BENTLEY, Design factors controlling the dynamic performance of instruments, *Trans. A.S.M.E.*, **62**, 421–32 (1940).
5. C. S. DRAPER and G. V. SCHLIESTETT, General principles of instrument analysis, *Instruments*, **12**, 137–42 (1939).
6. D. E. WEISS, Design and application of accelerometers, *Proc. Soc. Exp. Stress Anal.*, **4** (2), 89–102 (1946).
7. J. S. NISBET *et al.*, High frequency strain gauge and accelerometer calibration, *J. Acoust. Soc. Amer.*, **32** (1), 71–75.
8. I. VIGNESS, Shock motions and their measurement, *Exp. Mech.*, **1** (9), 13a–15a (1961).
9. S. J. LORING, Experimental determination of vibration characteristics of structures, *Proc. Am. Soc. Civil Eng.*, **73** (10), 1457–74 (1947).

XI

ANALOGUE METHODS FOR STRESS PROBLEMS

THE methods described in previous chapters have been concerned with the determination of stresses by observations on actual or model structures. It is sometimes convenient to study a stress problem experimentally by making measurements on a system whose behaviour is described by equations of the same kind as those which apply to the stress problem[1, 2]. The point can be illustrated[3] with reference to a damped spring–mass system and an electrical circuit, containing resistance, capacitance and inductance, fed from a source of alternating current. The equations describing the behaviour of these systems are as follows:

(*a*) Mechanical system:

$$M\frac{d^2x}{dt^2} + D\frac{dx}{dt} + Kx = F(t) \tag{11.1}$$

i.e. Inertia force + damping force + spring force = disturbing force

(*b*) Electrical system:

$$L\frac{d^2q}{dt^2} + R\frac{dq}{dt} + \frac{q}{c} = e(t) \tag{11.2}$$

where $q(t)$ = charge = $\int_0^t i\,dt$, i = current flowing in the circuit; i.e. voltage drop across inductance + voltage drop across resistance + voltage drop across condenser = supply voltage.

Equations (11.1) and (11.2) are mathematically similar and thus an analogy exists between the two systems, the following quantities being analogous:

Mechanical	*Electrical*
Mass (M)	Inductance (L)
Damping (D)	Resistance (R)
Spring force (K)	Reciprocal capacity ($1/C$)
Acceleration ($\mathrm{d}^2x/\mathrm{d}t^2$)	Rate of change of current ($\mathrm{d}^2q/\mathrm{d}t^2$)
Velocity ($\mathrm{d}x/\mathrm{d}t$)	Current ($\mathrm{d}q/\mathrm{d}t$)
Displacement (x)	Charge (q)
Force (F)	Voltage (e)

It would therefore be quite a straightforward matter to examine the behaviour of a lumped mass mechanical system by constructing an electrical analogue in which the various parameters of the mechanical system were represented by the corresponding electrical quantities. The advantages of doing this would be that the electrical circuit is very easily built, measurements in it are conveniently and reliably made by means of standard equipment and, finally, the parameters can be altered with very little difficulty. Even in the very simple mechanical system envisaged here, it would be very cumbersome to cover a wide range of mass, damping and spring force experimentally and the measurement of forces and displacements might call for quite elaborate instrumentation.

This illustration sums up the case for using analogue methods although of course the analogue equipment can itself become quite complicated. Although there are a few useful mechanical analogues in use, the convenience of electrical measurement and the development in electrical circuit theory have led to the evolution of electrical analogues for a whole range of engineering problems.

The design of electrical analogues is a large and specialised subject so that only a very brief introduction to analogue simulation will be attempted here. Before discussing electrical methods, however, mention will be made of the membrane analogy which is the most widely known and useful of the mechanical methods.

THE MEMBRANE ANALOGY

This analogy depends upon the fact that the differential equation expressing the lateral deflection of a uniformly stretched elastic membrane is the same as that expressing the shear stresses in a section subject to torsion[4, 5]. This is in fact Poisson's equation:

(*a*) For torsion of a prismatic bar:

$$\frac{\partial^2 \phi}{\partial x^2} + \frac{\partial^2 \phi}{\partial y^2} = -\,2G\theta \qquad (11.3)$$

in which ϕ is a torsional stress function, G is the modulus of rigidity and θ is the angle of twist, in radians, per unit length.

(*b*) For the lateral deflection of a stretched membrane:

$$\frac{\partial^2 z}{\partial x^2} + \frac{\partial^2 z}{\partial y^2} = \frac{q}{T} \qquad (11.4)$$

in which z is the deflection, q is the transverse force per unit area and T is the uniform tension per unit length at the boundary of the sheet. Comparison of (11.3) and (11.4) shows that the displacement represents the stress function if q/T is equated to $2G\,.\,\theta$.

In applying the method, the membrane may be either a uniformly stretched rubber sheet, a soap film or the meniscus surface which is formed between two immiscible liquids. Whatever is used for the membrane, a template is cut from a flat plate, with a hole in it representing the shape of the bar. The membrane is then stretched or formed across the hole and slight pressure applied to one side of it, as indicated in Fig. 11.1.

Under this condition it can be shown that:

a) The tangent to a contour line at any point of the deflected membrane gives the direction of the shearing stress at the corresponding point in the cross-section of the twisted bar.

(*b*) The maximum slope of the membrane at any point is proportional to the magnitude of the shearing stress at the corresponding point in the twisted bar.

(*c*) The volume included between the surface of the deflected membrane and the plane of the template is proportional to the torque of the twisted bar.

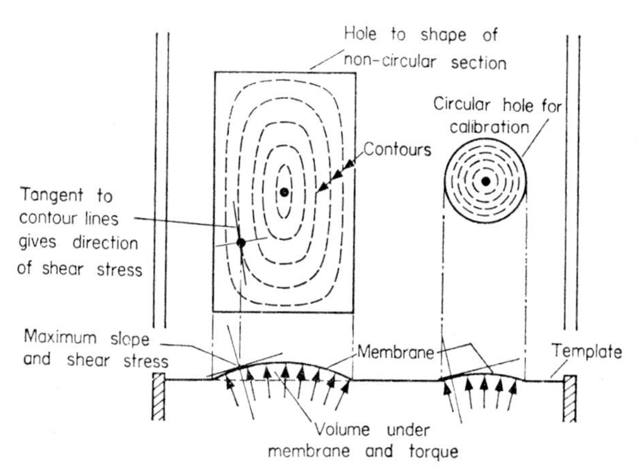

Fig. 11.1. The membrane analogy.

In order to calibrate the slopes, measured from the model, in terms of shear stress, a subsidiary experiment is conducted in which an identical membrane is stretched across a circular hole in the template and subjected to the same lateral pressure as the membrane over the hole representing the section being investigated; the maximum slope of the membrane covering the circular hole is then measured at some suitable point and the shearing stress at the corresponding point in a circular shaft is calculated. Since q/T is the same for both the circular and non-circular membranes, it follows that the stresses are compared for the condition that the angles of twist are the same in both cases. The ratio of the torques necessary to produce this angle of twist is found by calculating the ratio n of the volumes between the membranes and the plane of the template. Obviously this ratio gives the ratio of the torsional rigidities of the two sections. Suppose that S_0 is the shear stress at a point in the circular shaft and M_0 is the torque on this section; if M is the torque on the non-circular section and τ the shear stress at a particular point in that section. then:

$$\tau = \tau_0 \cdot \frac{S}{S_0} \tag{11.5}$$

where S_0 and S are the maximum slopes at the points considered in the circular and non-circular sections.

Various forms of experimental equipment have been devised for the membrane analogy depending on what membrane material is used. For a soap film membrane the apparatus shown in Fig. 11.2 devised originally by Griffith and Taylor[5], may be employed. In this, the membrane is formed over openings in a metal plate. Pressure is applied by displacing air from a closed flask into the cast-iron box; this is achieved by running soap solution from a burette into the flask. The height of the membrane at any point is determined by means of a micrometer depth gauge which is carried above the template on a loose sheet of glass, the measuring pointer of the gauge passing through a hole in the glass. The contours of the inflated soap film are plotted by setting the depth gauge to a particular reading above the surface of the template and moving it cautiously into contact with the membrane; the point of the micrometer is wetted with soap solution beforehand and contact is indicated when the film seems to move up to the point. The hinged drawing board is then swung down on a point on top of the micrometer to make a mark on a previously drawn outline of the section.

This procedure is repeated for as many points as necessary

to define the contour line. The volume enclosed under the soap film can be calculated from the contour plot by well known methods.

Various arrangements have been suggested for the measurement of the slope of the membrane, that devised by Quest

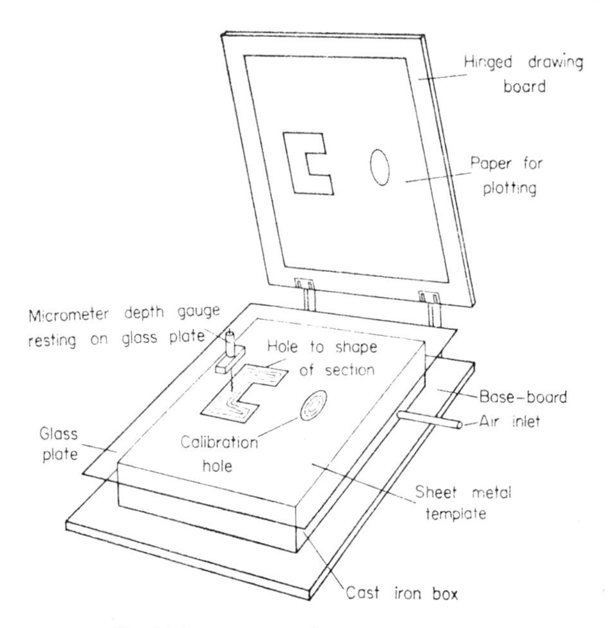

Fig. 11.2. Apparatus for soap film method.

being perhaps the most satisfactory from the point of view of accuracy and convenience. The arrangement of the instrument is shown in Fig. 11.3. A narrow beam of light is projected on to a graduated scale which is rotated about a vertical axis to interrupt the reflected beam. The reading of the arc gives 2α, equal to twice the slope of the membrane relative to the vertical; the azimuth, ϕ, gives the direction of the maximum slope relative to a convenient axis of reference. In Quest's apparatus the collimator was mounted on a fixed base and the membrane box was arranged on a milling machine table so that it could be traversed under the collimator beam and thus permit the determination of the slope at any point whose co-ordinates could be accurately determined from the table setting.

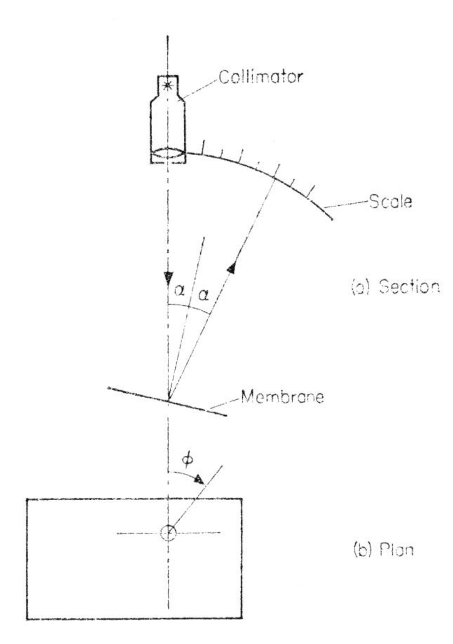

Fig. 11.3. Quest's collimator.

Various solutions have been used for forming the soap film, usually consisting of sodium or potassium oleate, glycerin and water. A typical formulation is 2 g sodium oleate, 30 cm^3 glycerin, 1 l. of water. The soap film is stretched over the holes with a smooth piece of celluloid or Perspex. Any excess solution must be drawn off with a straw or pipe cleaner. It is important to see that the pressure on both sides of the membrane is equal when the film is formed.

It is also possible to use a uniformly stretched rubber sheet in this experiment instead of a soap film. In this case, a grid is marked out on a sheet of surgical rubber which is then stretched in a suitable frame to give a uniform linear extension of one third. The stretched membrane is placed between a pair of templates, which define the shape of the cross-section of the member, and air pressure is applied from one side to produce the required transverse deflection. A rubber membrane can be made larger than a soap film and is, of course, less easily damaged so that height measurements are more easily determined.

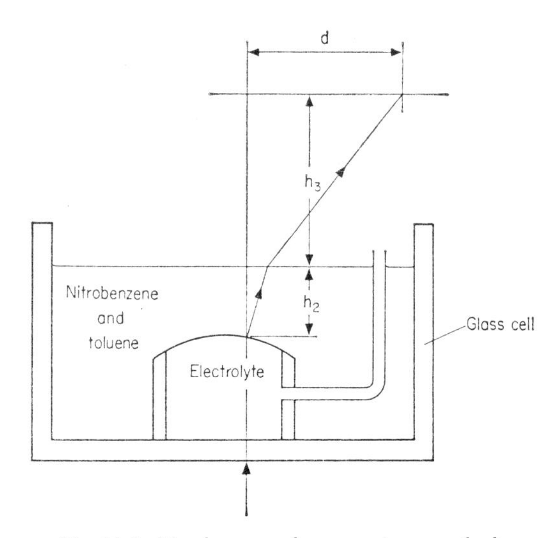

Fig. 11.4. Membrane analogy; meniscus method.

The third method of producing the equivalent of a uniformly stretched membrane is shown in Fig. 11.4. In this method a cell is erected on the base of a suitable flat bottomed glass tank to the shape of the cross section being investigated. This cell is filled with an electrolyte whilst the glass tank is filled with a mixture of nitrobenzene and toluene of the same density as the electrolyte. Capillary forces result in a constant tension on the surface and the meniscus behaves as a uniformly stretched membrane. The meniscus is distended by applying a slight excess pressure to the electrolyte by introducing additional liquid to the cell. Height contours can be determined by a micrometer screw gauge; contact of the pcinter with the surface of the electrolyte can be made to complete an electrical circuit which includes a small indicating light since the upper liquid is a non-conductor and the electrolyte permits the passage of a current. Optical methods have been devised for measuring the slope of the meniscus, one possibility being indicated in Fig. 11.4; here a ray of light is projected vertically through the bottom of the tank and through the liquids; the shift of the ray from its initial position is measured (d) and slope of the meniscus surface is given by:

$$\theta = \frac{d}{(1 + \mu_{21})(h_2 + \mu_{32} \,.\, h_3)} \tag{11.6}$$

where μ_{21}, μ_{32} are the indices of refraction upper mixture–electrolyte and air–upper mixture respectively.

Although not so well known as the soap film method, the meniscus method has a number of advantages: the meniscus surface is weightless and thus can be made larger than a soap film; the meniscus is undisturbed by air currents and lasts indefinitely, nor is it greatly affected by ambient temperature changes; finally, the slopes can be measured very accurately by optical methods.

In the torsion analogy the membrane method is used to represent Poisson's equation. It can also be used to represent physical systems described by Laplace's equation:

$$\frac{\partial^2 \phi}{\partial x^2} + \frac{\partial^2 \phi}{\partial y^2} = 0 \tag{11.7}$$

An important example in the field of stress analysis is $\phi = (\sigma_p + \sigma_q)$ the sum of the principal stresses in a two dimensional elastic stress field [(6)].

It can be shown that if transverse deflections are imposed on a uniformly stretched membrane, the deflected surface is given by:

$$\frac{\partial^2 z}{\partial x^2} + \frac{\partial^2 z}{\partial y^2} = 0 \qquad (11.8)$$

provided that the slopes and deflections of the sheet are small and that gravitational forces on the sheet are negligible. In this case no pressure is applied to the membrane.

Boundary conditions are imposed by erecting a template of the same shape as the plate being investigated and whose height is everywhere proportional to $(\sigma_p + \sigma_q)$. If a soap film is used, this is stretched over the template and measurements of the height of the film give $(\sigma_p + \sigma_q)$ at internal points. Corresponding techniques are possible using a rubber membrane or the meniscus between two immiscible liquids as in the torsion analogy.

This method is useful as a complement to photo-elasticity as the latter technique gives the boundary values of the principal stresses and also the difference of the principal stresses.

ELECTRICAL ANALOGY FOR LAPLACE'S AND POISSON'S EQUATIONS

Turning now to electrical methods, it is possible to show that the flow of electricity in a conducting sheet is represented by the Laplace equation. Considering an element $ABCD$ in such a sheet whose sides are δ_x and δ_y, if the currents flowing into the element are i_1, i_2, i_3, i_4 on the four sides, as shown in Fig. 11.5, and the resistivity of the plate is R ohms per unit square, then the voltage gradients across the four sides are:

$$\left.\begin{aligned} \left(\frac{\partial v}{\partial x}\right)_1 &= \frac{i_1 R}{\delta y} \qquad \left(\frac{\partial v}{\partial x}\right)_2 = \frac{i_2 R}{\delta y} \\ \left(\frac{\partial v}{\partial y}\right)_3 &= \frac{i_3 R}{\delta x} \qquad \left(\frac{\partial v}{\partial y}\right)_3 = \frac{i_4 R}{\delta x} \end{aligned}\right\} \qquad (11.9)$$

The rate of change of the voltage gradient is thus:

$$\left.\begin{aligned}\frac{\partial^2 v}{\partial x^2} &= \frac{(\partial v/\partial x)_2 - (\partial v/\partial x)_1}{\delta x} \quad \text{as} \quad \delta x \to 0 \\ \frac{\partial^2 v}{\partial y^2} &= \frac{(\partial v/\partial y)_4 - (\partial v/\partial y)_3}{\delta y} \quad \text{as} \quad \delta y \to 0\end{aligned}\right\} \tag{11.10}$$

Adding these two equations and substituting for the partial differentials from (11.8):

$$\frac{\partial^2 v}{\partial x^2} + \frac{\partial^2 v}{\partial y^2} = \frac{R}{\delta x \,.\, \delta y}(i_1 - i_2 + i_3 - i_4) \tag{11.11}$$

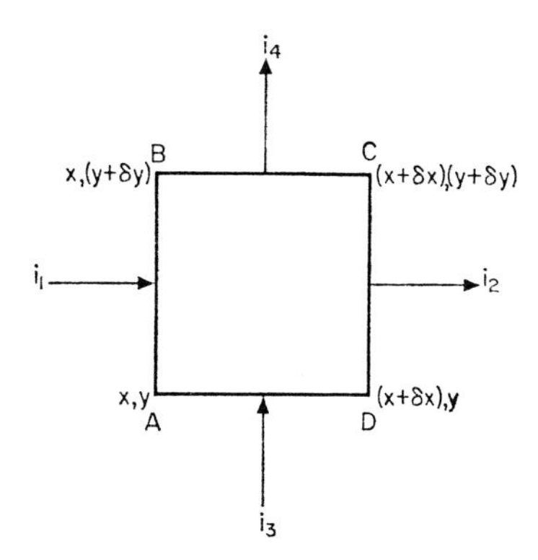

Fig. 11.5. Current flow in a conducting sheet.

It is known from Kirchoff's laws that the algebraic sum of the currents flowing into an element is zero, i.e. $(i_1 - i_2 + i_3 - i_4) = 0$ and thus:

$$\frac{\partial^2 v}{\partial x^2} + \frac{\partial^2 v}{\partial y^2} = 0 \tag{11.12}$$

In other words, the voltage distribution in a conducting sheet is described by Laplace's equation.

If another current is fed into the element, as indicated in

Fig. 11.6 the right-hand side of (11.10) is no longer equal to zero but is equal to $-Ri_i$ and thus:

$$\frac{\partial^2 v}{\partial x^2} + \frac{\partial^2 v}{\partial y^2} = -R_i \tag{11.13}$$

which is, of course, Poisson's equation.

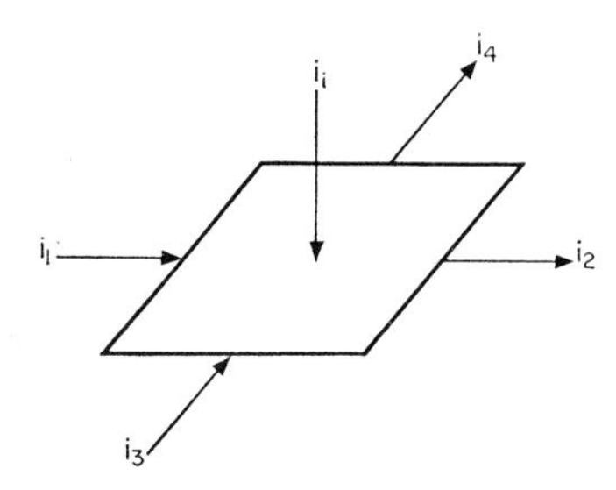

Fig. 11.6. Current flow in a conducting sheet with inflow to surface of element.

Thus by correctly applying the boundary conditions it is possible to represent by a conducting sheet analogy, a wide variety of physical systems which are described by these two equations. It is also possible to achieve the same result using a resistance network, small elements of the conducting sheet being "lumped up" into discrete resistors.

In the case of the Laplace equation[10] applicable to the determination of the sum of the principal stresses in a stressed plate, the procedure is to apply voltages to the edges of a conducting sheet proportional to the known boundary values; internal values are then obtained by measuring the voltage at as many points as may be required.

Setting up the boundary voltages presents certain practical difficulties since it is found impossible to obtain satisfactory results by applying potentials at discrete points. The following technique has, however, been devised by Stokey and Hughes and gives results of very good accuracy. A sheet of conducting paper is cut to the outline of the test piece, but with a margin of about 9 mm outside the true boundary. If the test piece is symmetrical the sheet may be cut along axes of symmetry, but in this case leaving no margin. A line 3 mm to 6 mm in width is then drawn round the boundaries about 1·5 mm from the line representing the true boundaries with resistance paint of the type used for printed electronic circuits. This line should have a resistance of 1000–4000 ohms/mm depending on whether voltage connections are widely or closely spaced. The actual value is not critical, but must be uniform and must be considerably lower than the resistance of the adjoining paper (which will be about 2000 ohms square). It will be noted that edges lying along axes of symmetry are not painted in this way, and no voltage connections are made to them because no current will flow across such lines. Boundaries which are at constant potential are edged with a high conductivity paint.

Having prepared the conducting sheet in this way, the next step is to apply the necessary boundary voltages from a simple voltage divider. This may consist of a wire 3 m in length having a resistance of about 30 ohms. Connections are tapped off at appropriate points along the wire with crocodile clips and connected to the boundary line with silver conducting paint. With resistances of the order mentioned above, the connection of successive clips will have no appreciable effect on those already set. The spacing of connections on the boundary is related to the gradient of the boundary voltage; if this is small, connections may be 50–75 mm apart, reducing to, say, 3 mm where the boundary stresses are changing rapidly.

In order that measurements of the voltages at internal points should not affect the potential distribution established by the boundary conditions, it is essential to employ a device with a very high impedance. A convenient method is to connect an electronic voltmeter or an oscilloscope with a d.c. amplifier to the point on the voltage divider at which the required voltage exists. An alternative is to use a calibrated potentiometer, as shown in Fig. 11.7.

Internal voltages can be measured either at points on a reference grid or alternatively contours can be traced on the paper. These will be equivalent to lines of constant $(\sigma_p + \sigma_q)$ in the stress field, known as isopachic lines.

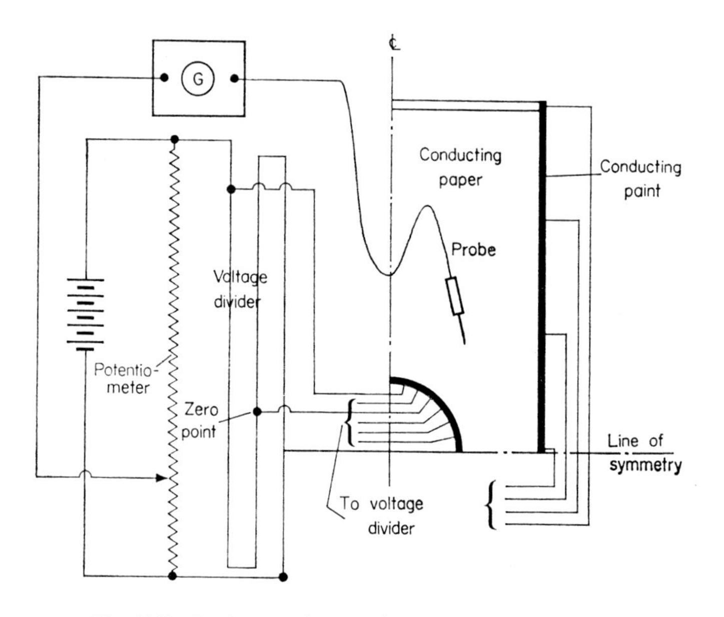

Fig. 11.7. Conducting sheet analogy for Laplace equation.

Examination of Poisson's equation shows that a certain constant current has to be fed in to the surface of the sheet, the boundaries being kept at a constant voltage. This current corresponds to the transverse pressure applied in the soap film analogy, the flat boundary being equivalent to constant voltage round the perimeter of the conducting sheet. Feeding in a uniform current over the surface of the sheet presents an obvious difficulty and although it is possible to feed in current at discrete points[11], it is more convenient to transform the torsion equation to a Laplace equation and then to carry out the experiment as described above for the $(P + Q)$ analogy. This transformation is readily achieved, as first demonstrated by Griffith and Taylor, by replacing ϕ in equation (11.3) by:

$$\psi = \phi + \frac{G\theta}{2}(x^2 + y^2) \tag{11.14}$$

The original equation then becomes:

$$\frac{\partial^2\psi}{\partial x^2} + \frac{\partial^2\psi}{\partial y^2} = 0 \tag{11.15}$$

and the boundary potential will now be:

$$\psi_B = \frac{G\theta}{2}(x_b^2 + y_b^2) \tag{11.16}$$

that is, instead of maintaining zero potential round the perimeter of the sheet, the potential at x_B, y_B, is a function of these co-ordinates as defined by (11.15). Internal potentials then give values of ψ and the corresponding values of ϕ are obtained from equation (11.14). As in the soap film analogy, the maximum shearing stress at a point is given by the maximum slope of the ϕ function, i.e. $\partial\phi/\partial n$ where n is the direction normal to the ϕ contours.

Suitable material for the conducting sheet is available commercially as Teledeltos paper. This can be obtained in rolls of 450 mm width so that measurements can be taken on sheets of reasonably large size. There is, however, a tendency for the resistance of this paper to be about 10 per cent different along the grain as compared with that measured across the sheet; this should be checked before cutting the outline and if necessary the dimensions in the direction of greatest resistance reduced to compensate for the difference in resistance. Conducting and resistance paints are also available commercially.

An alternative to conducting paper in this analogue is the use of an electrolytic tank; the principle here is the same, a tank containing a conducting fluid replacing the conducting sheet.

RESISTANCE NETWORK ANALOGUES

The possibility of using a network of discrete resistors instead of a conducting sheet has already been mentioned. It is also possible to devise resistance networks to represent a variety of elastic stress–strain and structural systems [13, 14].

One of the most important problems in engineering is that of determining the stresses in plates and slabs. In the case of plates loaded in their own plane, the state of stress is described by the biharmonic equation:

$$\nabla^4\phi = \frac{\partial^4\phi}{\partial x^4} + 2\frac{\partial^4\phi}{\partial x^2\,\partial y^2} + \frac{\partial^4\phi}{\partial y^4} = 0 \qquad (11.17)$$

in which ϕ is the Airy stress function such that the state of stress at a point is defined by:

$$\sigma_x = \frac{\partial^2\phi}{\partial y^2} \qquad \sigma_y = \frac{\partial^2\phi}{\partial x^2} \qquad \tau_{xy} = \frac{\partial^2\phi}{\partial x\,\partial y}$$

It has been shown [13] that the biharmonic equation can be represented by a cascade arrangement of rectangular resistance networks interconnected at their nodes, as indicated in Fig. 11.8. The potentials at the nodes of the upper network represent a stress function:

$$Y = \left(\frac{\partial^2\phi}{\partial x^2} + \frac{\partial^2\phi}{\partial y^2}\right) \qquad (11.19)$$

whose value along the boundaries of the network must be known. The node voltages of the lower network are to represent the required function ϕ; this is achieved by connecting the corresponding node points through resistors, r, which are much larger than those making up the networks. The current through the resistor r is then:

$$I_0' = \frac{Y_0 - \phi_0}{r} \qquad (11.20)$$

and since r is large, $Y_0 \gg \phi_0$ so that:

$$I_0' \simeq \frac{Y_0}{r} \tag{11.21}$$

The voltages at the nodes of the lower network are then defined by the Poisson equation:

$$\nabla^2\phi = K \cdot \frac{Y}{r} \tag{11.22}$$

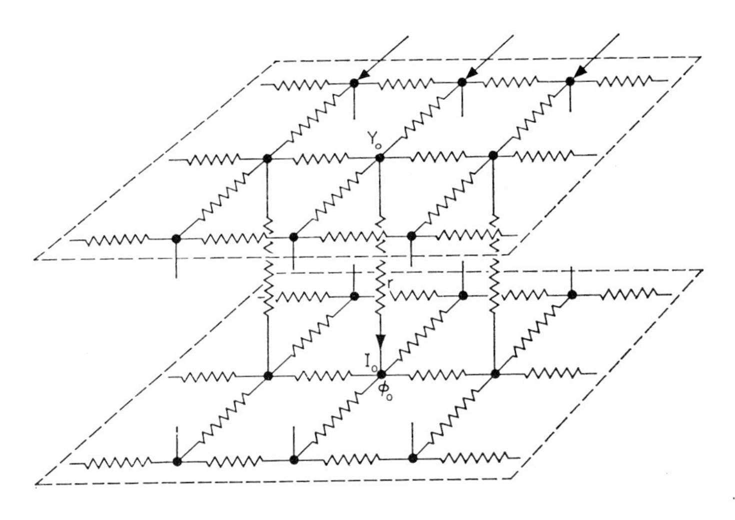

Fig. 11.8. Network analogue for a plate stressed in its own plane.

But since the voltage distribution Y approximately satisfies a Laplace equation:

$$\nabla^2(\nabla^2\phi) = \nabla^4\phi = 0 \tag{11.23}$$

and thus the node voltages in the lower network represent the required stress function. Methods are available for calculating the boundary values of the Airy stress function so that once these are known, the necessary boundary voltages can be set on the lower network. Voltages applied to the upper network have to be in accordance with the applied boundary loading.

As the two networks interact, the simultaneous establishment of the two sets of boundary values requires a process of successive correction.

The equation for the deflection of transversely loaded plates is:

$$\frac{\partial^4 \omega}{\partial x^4} + 2\frac{\partial^4 \omega}{\partial x^2\, \partial y^2} + \frac{\partial^4 \omega}{\partial y^4} = \frac{q}{D} \tag{11.24}$$

$$\text{i.e.} \quad \nabla^4 \omega = \frac{q}{D} \tag{11.25}$$

where q is the load per unit area and $D = \frac{Eh^3}{12} \Big/ (1 - \nu^2)$, h being the thickness of the plate. It is clear that there is the same sort of relationship between the biharmonic equations for plates loaded in their own plane and that for plates loaded tranversely as exists between the Laplace and Poisson equations. Thus a network analogue for (11.24) can be developed from that representing (11.17) by feeding in current to the internal node points of the upper network proportional to the intensity of loading on the transversely loaded plate. It is of course evident that these network solutions are approximations to the extent that a continuous plate is being represented by a finite mesh of resistances.

Resistance networks can also be designed to simulate beams and frameworks, one of the most interesting being a representation of the slope–deflection equation which can be used for the analysis of rigid frame structures (15). Referring to the member AB shown in Fig. 11.9(a) the slope–deflection equations are:

$$\left.\begin{aligned} M^F{}_{AB} &= m_{AB} - \frac{2EI}{L}(2\theta_A + \theta_B) \\ M^F{}_{BA} &= m_{BA} - \frac{2EI}{L}(2\theta_B + \theta_A) \end{aligned}\right\} \tag{11.26}$$

where m_{AB} and m_{BA} are the fixed end moments for the member. For the electrical circuit shown in Fig. 11.9(b) it can be established by applying Kirchoff's rules that:

$$\left.\begin{aligned} I_{AB} &= i_{AB} - \frac{1}{R}(2V_A - V_B) \\ I_{BA} &= i_{BA} - \frac{1}{R}(2V_B - V_A) \end{aligned}\right\} \qquad (11.27)$$

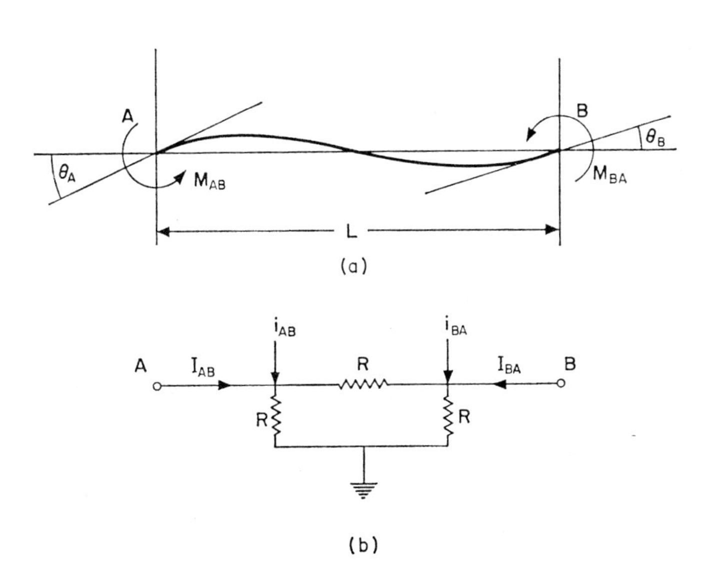

Fig. 11.9. Analogous circuit for slope-deflection equation.

These equations are the same as (11.26) apart from a difference of sign. It can be seen that the following quantities correspond:

Network current (I)	End moments in member (M)
Feed current (i)	Fixed end moments (m)
Voltage (V)	Angle of rotation (O)
Resistance (R)	$L/2EI$.

In order that the electrical circuit should be analogous the following relationships must therefore be established:

$$
\begin{aligned}
I_{AB} &= +p \,.\, M_{AB} & I_{BA} &= -pM_{BA} \\
i_{AB} &= +p \,.\, m_{AB} & i_{BA} &= -pm_{AB} \\
V_A &= +q \,.\, \theta_A & V_B &= -q \,.\, \theta_B \qquad (11.28) \\
R &= \frac{q}{p} \cdot \frac{L}{2EI}
\end{aligned}
$$

where p and q are convenient factors. Substitution of these values in (11.27) will be found to yield the original slope–deflection equation (11.26). At the nodes of a framework, the sum of the member end moments must be zero and the joint

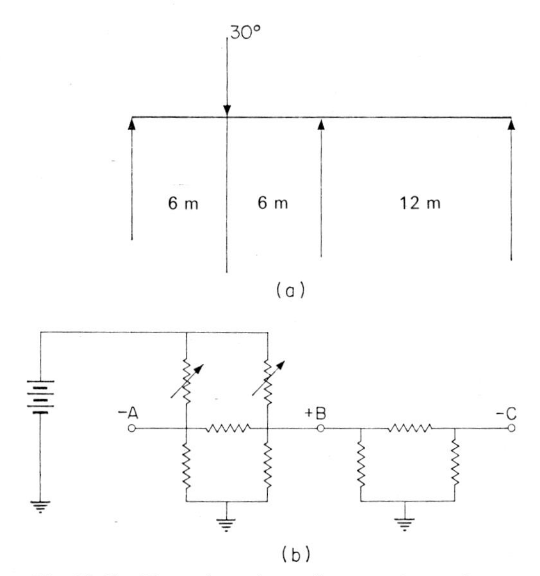

Fig. 11.10. Network analogue for a continuous beam.

rotation must have a unique value; in the analogue the equivalent conditions are that the total current flowing into a joint must be zero and that there must be a single voltage at the joint. It is also essential to specify a sign convention and it is convenient to treat *clockwise moments* and *currents directed towards* the node as *positive*. Inspection of the expressions in (11.28) shows that positive signs occur in all equations concerning A and negative signs in all concerning B, so that alternate + and − signs are attached to the nodes of the network. The following rules must then be observed:

Sign of feed current = (sign of f.e.m.) × (sign of node)
Sign of fixing moment = (sign of network current) × (sign of node)
Sign of rotation = (sign of voltage) × (sign of node).

As a simple illustration of the application of the analogue, we may consider the two-span continuous beam shown in Fig. 11.10. The dimensions, fixed end moments and the necessary electrical parameters are as follows:

	A		B		C
Span (m)		12		12	
I (m^4)		36		36	
F.E.M. (kN m)	−462		+462		0

Scales p: 1 mA ≡ 30 kN m $q = \frac{2EIR}{L} \cdot p = 1800$ V/rad

	A		B		C
Signs	−		+		−
$R\left(\propto \frac{L}{I}\right)$		1000		1000	
Feed current (mA)	+15		+15		

The circuit is set up as shown in Fig. 11.10(b). It will be noted that as the moments at A and C are zero, the currents at these points are zero, and they are left unconnected (if the end of a beam is fixed, this is represented by earthing the corresponding point in the network, zero rotation being represented by zero voltage). With the circuit set up as in Fig. 11.10(b) and the correct feed currents applied, the network currents and voltages are measured and converted to moments by applying the scale factors. The results obtained are as follows:

	A	*B*		*C*
Current (mA)	0	+11·25	−11·25	0
Moment (kN m)	0	+337·5	−337·5	0
Voltage	+11·25	+7·5		+3·75
Angle of station (rad)	0·00624	+0·00416		−0·00208

Having found the support moment at *B*, it is a simple matter to draw the bending moment diagram for the whole beam by superimposing the free moment and fixing moment diagrams.

Continuous beams of any number of spans and having any combination of support conditions can be dealt with in this way. Rigid frames can also be represented and analysed by connecting together groups of resistors representing the members, provided that there is no sway, the moments at the joints are obtained by current measurements as in continuous beams. If, however, there is sway, auxiliary, shear circuits have to be introduced to reproduce this effect. The basis of the sway correction procedure may be understood by first considering the effect of relative lateral displacement of the ends of a member. Referring to Fig. 11.11, the slope–deflection equations for a member, allowing for lateral displacement are:

$$\left.\begin{aligned} M_{AB} &= m_{AB} - \frac{2EI}{L}\left(2\theta_A + \theta_B + \frac{3\Delta}{L}\right) \\ M_{BA} &= m_{BA} - \frac{2EI}{L}\left(2\theta_B + \theta_A + \frac{3\Delta}{L}\right) \end{aligned}\right\} \quad (11.29)$$

The corresponding electrical circuit equations are:

$$\left.\begin{aligned} I_{AB} &= i_{AB} - \frac{1}{R}(2V_A - V_B - v) \\ I_{BA} &= i_{BA} - \frac{1}{R}(2V_B - V_A + v) \end{aligned}\right\} \quad (11.30)$$

Thus we may represent relative lateral displacement of the ends of a member by introducing terminal voltages $\pm v$ proportioned to $3\Delta/L$. The voltage scale is, as before, q, so that:

$$v = q \cdot \frac{3\Delta}{L} \tag{11.31}$$

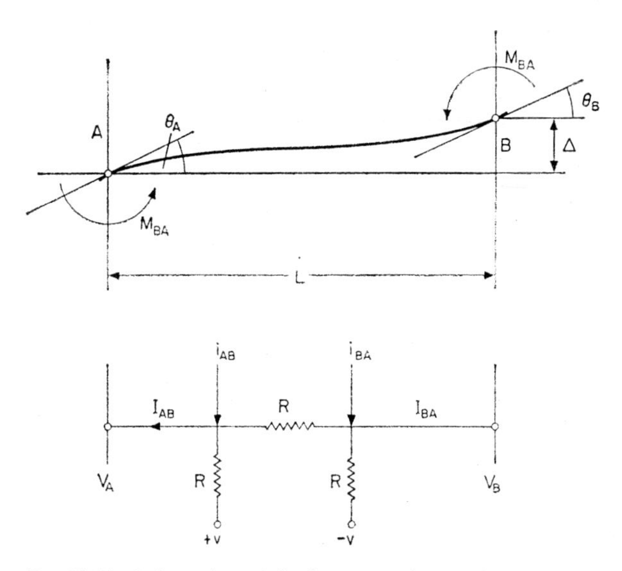

Fig. 11.11. Relative lateral displacement of ends of a member.

We can, therefore, apply the equivalent of horizontal sway to each storey of a network representing a frame by introducing equal and opposite voltages v to the ends of each column. The voltages will be the same for each storey, but will differ from one storey to another. The principle is in effect the same as in moment distribution: the analysis is first carried out under the no-sway condition and the results are adjusted to eliminate the horizontal forces which would be necessary to prevent lateral movement of the frame. The method of achieving this may be explained with reference to the frame shown in

Fig. 11.12(a). This structure is represented by the circuit shown in Fig. 11.12(b) for the no-sway condition and by the modified circuit of Fig. 11.12(c) for the analysis with sway. The auxiliary shear circuits are shown at the right of the main network; the leads *a*, *b*, *c*, from each column are connected

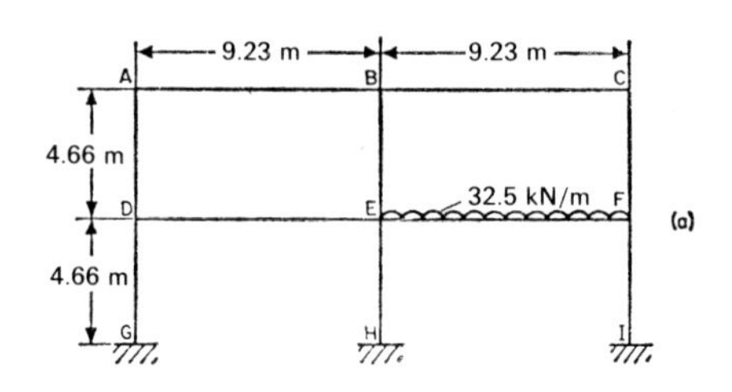

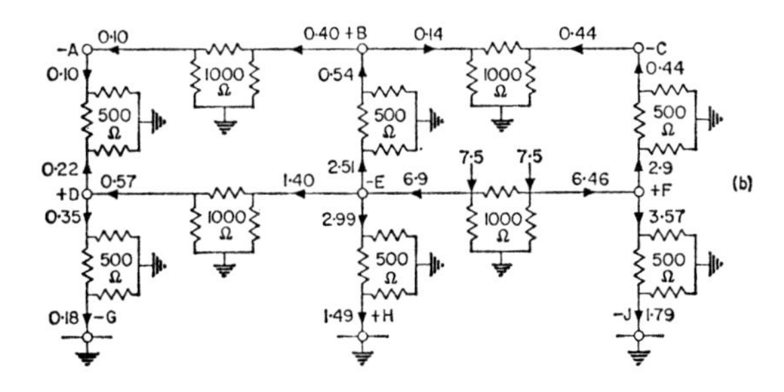

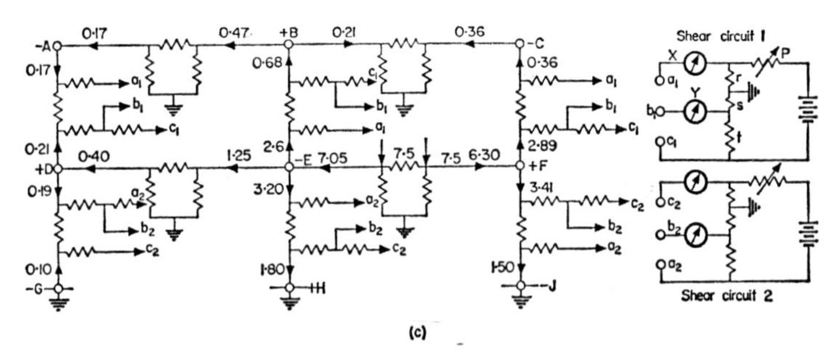

Fig. 11.12. Analysis of a rigid frame by Bray's resistance network method.

to the corresponding terminals of the shear circuit for each storey. The effect of this shear circuit is to introduce equal but opposite voltages at the ends of each of the column lengths *AD*, *BE*, and *CF*, giving the effect of lateral displacement or sway of the storey. The magnitude of the sway voltage is controlled by the resistance *P*. To ensure that the impressed voltages are equal without appreciable error, it is necessary that the values of resistors *r*, *s*, and *t* should be small compared with those of the main network. The second requirement of the shear circuit is that it should provide means of sensing the propping forces. This is achieved by the circuit arrangements shown; by applying Kirchoff's rules to the circuit, it can be demonstrated that the readings of milliameters *X* and *Y* are as follows:

$$\left.\begin{aligned}
&\text{Meter } X\text{: } \frac{2}{3}\left(I_{AD} + I_{BE} + I_{CF}\right) + \frac{1}{3}\left(I_{DA} + \right.\\
&\qquad\qquad\left. + I_{BE} + I_{FC}\right) + \frac{2v}{R}\\
&\text{Meter } Y\text{: } \frac{2}{3}\left(I_{DA} + I_{EB} + I_{FC}\right) + \frac{1}{3}\left(I_{AD} + \right.\\
&\qquad\qquad\left. + I_{EB} + I_{CF}\right) - \frac{2v}{R}
\end{aligned}\right\} \quad (11.32)$$

The difference of the readings of the two meters is therefore:

$$\delta I = \frac{1}{3}\left(I_{AD} + I_{EB} + I_{CF}\right) - \frac{1}{3}\left(I_{DA} + I_{BE} + I_{FC}\right)$$

or in terms of moments:

$$\delta I = \frac{p}{3}\left(M_{AD} + M_{EB} + M_{CF} + M_{DA} + M_{BE} + M_{FC}\right)$$

$$= \frac{p}{3} \cdot SL$$

where *P* is the moment–current scale factor;
S is the horizontal shear in the storey;
L is the length of the columns.

That is, the difference of the readings of the two meters indicates the storey shear or propping force.

In an experiment, the procedure is to increase the setting of P so that practically no current flows into the network from the shear circuit and the analogue simulates the no-sway condition. The storey shears are then observed from the shear circuit meters and the resistances P are adjusted iteratively until these shears are brought to zero.

Essentially the same method can be used to investigate wind loads and settlement stresses in a frame; modifications have been devised to deal with members of variable section and to simulate semi-rigid connections and plastic hinges. The analogue is thus very flexible, the equipment simple and its operation rapid; these advantages combine to make this analogue a most valuable experimental tool.

A discussion of the many other electrical analogues, using alternating current networks, is beyond the scope of this book. It may be mentioned, however, that these analogues can be devised for the investigation of vibration and transient force problems in beams and plates. Details of these methods will be found in references 1 and 16–19.

BIBLIOGRAPHY

1. W. J. Karplus, *Analog Simulation.* McGraw-Hill, 1958. (A general treatise on electrical analogues).
2. J. Pérès, *Les Méthodes d'anologie en méchanique appliquée*, Proc. 5th Inst. Congress App. Mech., pp. 9–19. J. Wiley & Sons, 1939.
3. Th. von Karman and M. A. Biot, *Mathematical Methods in Engineering*, pp. 228–233, 372–74. McGraw-Hill, 1940. (For analogy between mechanical and electrical oscillations.)
4. S. Timoshenko, *Strength of Materials*, pp. 266–274. Macmillan, 1941. (Membrane analogy for torsion.)
5. A. A. Griffith and G. I. Taylor, The use of soap films in solving torsion problems, *Proc. Inst. Mech. Eng.*, 755–809 (1918).
6. J. G. McGivern and H. L. Supper, A membrane analogy supplementing photo-elasticity, *J. Franklin Inst.*, **217**, 491–504 (1934).

7. T. J. Higgins, Analogic experimental methods in stress analysis as exemplified by Saint Venant's torsion problem, *Proc. Soc. Exp. Stress Anal.*, **2** (2), 17–27 (1944).
8. H. M. Westergaard, Graphostatics of stress functions, *Trans. A.S.M.E.*, **56**, No. 3, 144–50 (1934).
9. G. Liebmann, Electrical analogues, *Brit. J. App. Phys.*, **4**, 193 (1953).
10. W. F. Stokey and W. F. Hughes, Tests of conducting paper analogy for determining isopachic lines, *Exp. Stress Anal.*, **12** (2), 77–82 (1955).
11. N. S. Waner and W. W. Soroka, Stress concentrations for structural angles in torsion by the conducting sheet analogy, *Proc. Soc. Exp. Stress Anal.*, **11** (1), 19–26 (1953).
12. N. E. Friedmann *et al.*, Solution of torsional problems with the aid of the electrical conducting sheet analogy, *Proc. Soc. Exp. Stress Anal.*, **13** (2), 1–6 (1956).
13. P. J. Palmer and S. C. Redshaw, Experiments with an electric analogue for the extension and flexure of flat plates, *Aeronautical Quart.*, **6**, 13–30 (1955).
14. G. Liebmann, Solution of plane stress problems by an electrical analogue method, *Brit. J. App. Phys.*, **6**, No. 5, 145–57 (1955).
15. J. W. Bray, An electrical analyser for rigid frameworks, *Structural Engineer*, **35** (8), 297–311 (1957).
16. G. W. Riesz and B. J. Swain, Structural analysis by electrical analogy, *Exp. Stress Anal.*, **12** (1), 13–27 (1954).
17. G. Brouwer and S. van Der Meer, A network analogy of a statically loaded two dimensional frame, *Exp. Stress Anal.*, **15** (1), 35–42 (1957).
18. R. H. Scanlan, Resistance network solution of some structural problems in deflection and stability, *Exp. Stress Anal.*, **16** (1), 117–128 (1958).
19. N. L. Svensson, An electric analogue for the limit analysis of framed structures, *Struct. Engineer*, **37** (10), 292–8 (1959).

XII

SELECTION OF METHOD OF ANALYSIS

ALTHOUGH there are usually several possible methods available for carrying out an experimental investigation of stresses in a structure or element it is important to select the technique to be used with great care so as to obtain the most satisfactory results with the minimum expenditure of time, effort and money. With experience this can be done fairly easily, but for the benefit of students approaching the subject for the first time, a particularised discussion, supplemented by illustrative examples, may be useful.

On embarking upon an investigation, the items of information listed in Table 12.1 will first have to be assembled. It is essential at the outset to define the purpose of the investigation as the measurements taken will to a large extent be influenced by the use to be made of the experimental results. For example, an analysis for design purposes may be required to provide values of maximum stress at critical points and sections whereas a research project may call for a series of detailed stress analyses on models differing in only one particular. Site tests are often limited to deflection measurements. Items 3–10 all have an important bearing on the selection of the method of analysis whilst 11 is an essential, and possibly over-riding, consideration which must be taken into account in any practical situation. Having gathered and assessed all this information decisions must be reached on the method of analysis to be used, the method of loading (except in the case of analogue methods) and, where appropriate, the choice of material.

These decisions are reached by relating the primary information to the characteristics of the various methods of

Table 12.1
Selection of Method of Experimental Stress Analysis

	Primary information required for outline of problem
1. Purpose of investigation?	e.g. Design, basic research, site test, etc.
2. Extent of investigation?	e.g. Complete analysis, surface stresses, stresses across sections, etc.
3. Site or laboratory investigation?	If site investigation, note special considerations such as ambient temperature, need for waterproofing, accessibility of gauge points, etc.
4. Type and dimensions of structure, element or detail?	Note need for three dimensional analysis or other special factors.
5. Material of prototype?	Consider special properties, e.g. whether elastic or not, creep characteristics, dimensional stability with changes in moisture content, etc.
6. Measurements required: elastic or plastic strains?	Ascertain whether elastic stress analysis adequate or whether behaviour up to failure to be studied.
7. Magnitude of stresses and strains in prototype?	Assess the orders of magnitude of stresses and strains which will have to be measured.
8. Nature of loading and magnitude of forces?	Static, dynamic, gravitational, etc.
9. Period over which readings to be taken?	Consider need for long term stability of measuring system.
10. Availability of equipment?	Ascertain what equipment and supporting facilities are at hand. Consider availability or otherwise of technical assistance and workshop facilities.
11. Time and money available?	Make an assessment of the time and the financial resources likely to be available for the investigation.

analysis. Having made a provisional selection of method the questions of loading and material can then be considered. In the case of investigations on actual structures the material, and possibly the loading, will of course be predetermined and will probably limit the choice of method of analysis. Similarly if creep, moisture and thermal effects are important, the choice of material for a model analysis and thus the method to be used, may thereby be restricted. When a choice of several methods presents itself, the final selection may be controlled by considerations of available facilities, time and cost. These factors are also likely to have a bearing on the selection of scale for a model study. It is usually advantageous to make a model scale as large as possible for accuracy and convenience of measurement, but there are likely to be limits as regards the size of models of large structures, such as dams, which can be accommodated in a laboratory; cost will also be a limiting factor in such cases not only as regards the model itself, but also in relation to the necessary loading gear.

The methods of analysis which have been discussed in this book are listed in Table 12.2 along with a few notes concerning their fields of application. Tables 12.3 and 12.4 contain respectively, similar summaries of loading methods and materials.

ILLUSTRATIVE EXAMPLES

We may now consider a number of examples of investigations which have actually been carried out to illustrate the selection of method of analysis in a variety of problems.

Stress Distribution in Portal Frame Connections[1]

Outline of Problem (referring to Table 12.1):

1. Purpose: research
2. Extent of investigation: complete analysis
3. Laboratory investigation

Table 12.2
Methods of Analysis

	Indications regarding applications
1. *Strain Gauges*	
(a) Mechanical	Few points: Static loading. Relatively large strains. Surface mounting. Gauge points must be accessible.
(b) Pneumatic	Limited number of points. Static loading. High sensitivity possible. Internal mounting possible. Centralised reading.
(c) Electrical resistance	Large number of points. Static or dynamic loading. High sensitivity. Internal mounting possible. Centralised reading. Recording possible. Very small gauges available.
(d) Electrical inductance	Possess most characteristics of resistance strain gauge but transducers are larger. High inherent stability and high power output. Large range of strain measurable.
(e) Electrical capacitance	More limited application but useful for measuring dynamic deflections of light members.
(f) Vibrating wire	Large number of points. Static loading. High sensitivity and stability. Internal mounting possible, e.g., in concrete members. Centralised reading. Wide range of gauge lengths possible.
2. *Surface Coatings*	
(a) Brittle lacquer	Useful for preliminary investigation and as supplement to strain gauging.
(b) Photo-elastic coatings	Surface stresses on any shape of object. Large strains can be measured. Static or dynamic loading.
3. *Photo-elasticity*	
(a) Normal two dimensional	Complete exploration of stresses in plates and beams. Dynamic loading possible.
(b) Frozen stress	Analysis of three dimensional systems, as in dams.
4. *Structural Model Analysis*	
(a) Direct methods: Strain gauges	Electrical resistance gauges.

Table 12.2 (contd.)
Methods of Analysis (contd.)

	Indications regarding applications
Photo-elasticity	Useful for a analysis of geometrically complex structures such as dams.
Moiré fringe method	For analysis of slabs.
Moment indicator	For rigid frames.
(b) Indirect methods:	
Beggs deformeter	
Large deformation methods	Applicable to most types of redundant structures.
5. *Analogues*	
(a) Membrane analogy	For solution of Laplace and Poisson's equations, e.g. for $(\sigma_p + \sigma_q)$ and for torsion problems.
(b) Conducting paper analogue	Applicable to same problems as membrane analogy.
(c) Resistance networks	Simple networks can represent Laplace and Poisson's equations. Double network can represent biharmonic equation for plates and slabs. Network available for slope–deflection equation.

Note—More complicated electrical analogues are available for study of dynamic problems.

Table 12.3
Methods of Loading

	Remarks
I. *Application of Load*	
A. *Static or Slowly Varying*	
(a) Dead load	Accurate and simple but limited by space requirements.
(b) Dead load with levers	Useful for maintaining constant load.
(c) Screw jacks	Limited application.
(d) Hydraulic jacks	Very flexible. Simultaneous multi-point loading possible. Control from central point.
(e) Pneumatic devices	Occasionally useful in special cases but store large amount of energy thus to be used with care.
(f) Centrifugal loading	Simulation of gravitational acceleration.

Table 12.3 (contd.)

Methods of Loading (contd.)

	Remarks
B. *Dynamic*	
(a) Mechanical inertia	Vibration exciters and impact loading.
(b) Hydraulic jacks	Cyclic loading at a number of points.
(c) Electro-magnets	Vibration exciters.
II. *Load Measurement*	
(a) Dead load	Direct mass standard.
(b) Elastic devices	Wide range of possible devices. Can be arranged to give electrical signal for recording or distant reading. Generally most useful class of load measuring instruments.
(c) Piezo-electric devices	Pressure sensitive.
(d) Photo-elastic balances	Special applications in small scale experiments.

Table 12.4

Materials

	Remarks
1. *Metals*	
(a) Steel	High modulus of elasticity.
(b) Aluminium alloy	Lower modulus. Easy to machine but difficult to join.
(c) Brass	Easy to machine and to join.
2. *Cement and Plaster*	
(a) Cement mortar	High alumina cement reduces time necessary to produce models.
(b) Plaster	Very rapid preparation of models.
3. *Plastic*	
(a) Epoxy resin	Can be cast and easily jointed (Araldite).
(b) Methyl methacrylate	Available as sheets and tubes, etc. Easily cut and jointed (Perspex).
(c) Polyethylene	Can be cast. Very low modulus (Alkathene).

Note—This list contains only a selection of the more commonly used materials.

4. Portal frame connections between two I-section members. Variety of shapes to be considered
5. Prototype—mild steel
6. Elastic analysis
7. Not relevant: model analysis contemplated
8. Small loads, dependent on material used for models
9. Short term measurements
10. All methods considered
11. Strict economy—an important consideration.

From this list, we may note that a complete elastic analysis of a structural detail was required and that a variety of alternative shapes had to be considered. In view of these requirements, and of the need for economy, photo-elasticity was clearly indicated. One might also have considered surface coating methods or resistance strain gauges (or a combination of both) on reduced size steel connections, but as a number of possible details had to be investigated and taking considerations of cost into account there is little doubt that the most appropriate method was photo-elasticity. It was clearly desirable, however, that there should be a limited number of control tests on steel frames in order to check the applicability of the photo-elastic results. These tests were conveniently carried out with resistance strain gauges.

For photo-elasticity the most suitable material available at the time was "Catalin" but an epoxy resin would now be used instead. Load was applied either by dead load or by a proving ring and screw gear.

As the members were of I-section it was necessary either to build them up by glueing suitable pieces together after machining to shape or to machine them out of a thick plate. The latter procedure was adopted in this case. It was found difficult to preserve geometrical similarity between model and prototype as regards flange dimensions and web thickness; the latter had to be made proportionally thicker in the model in order to produce a measurable fringe pattern. It was found, however, that this difficulty could be overcome and that the stress system in the prototype was reasonably well represented.

Fig. 12.1. Results of tests in portal frame connections.

Stress Distribution in a Welded Steel Dock Gate[(2)]

Outline of Problem:

1. Purpose: Research
2. Extent of investigation: Determination of maximum stresses in main members
3. Site test. Gauges to be submersible. Access to gauge stations rather difficult

4. Mitred leaf dock gate; 14·5 m dock opening
5. Welded mild steel
6. Elastic stresses
7. Stresses of up to 30 N/mm^2 expected
8. Hydrostatic loading; slowly varying with tide
9. Readings to extend over several weeks
10. All methods available
11. No special restrictions as regards time—reasonable economy to be exercised.

The main points to note in this investigation are the rather difficult conditions under which the tests had to be carried out, requiring completely waterproof gauges which would be stable over a prolonged period. Inaccessibility of the gauge stations ruled out demountable gauges of the "Demec" type and conditions were not very suitable for resistance strain gauges. Vibrating wire gauges were selected as they met the various requirements. Inductance gauges would also have been suitable, but the vibrating wire type was used in order to benefit from the experience of other laboratories in using these gauges under somewhat similar conditions.

The gauges were of the type shown in Fig. 5.13 and can be seen at various points on the gate in Fig. 12.2.

Stress Distribution in a Diamond Head Buttress Dam [3]

Outline of Problem:

1. Purpose: Information for design
2. Extent of investigation: as complete an analysis as practicable particularly in the buttress head
3. Laboratory investigation
4. Diamond head buttress dam (see Fig. 12.3)
5. Prototype—concrete
6. Elastic analysis
7. Arbitrary stress scale to suit method of analysis
8. Gravitational and hydrostatic. Load scale factors arbitrary
9. Short term tests

10. All methods considered but available laboratory space limited
11. Reasonable economy to be exercised. Limited time available for investigations.

Fig. 12.2. Dock gate test.

The main problems which presented themselves in this study were:

(*a*) the need to examine the three dimensional stress systems in the diamond head;

(*b*) the simulation of gravitational and hydrostatic loading conditions; and

(*c*) the desire to examine in detail the stress distribution in the buttress head, particularly around vertical holes in a special buttress accommodating the ground sluice of the dam.

In this case there was little difficulty in arriving at the conclusion that frozen stress photo-elasticity was the only

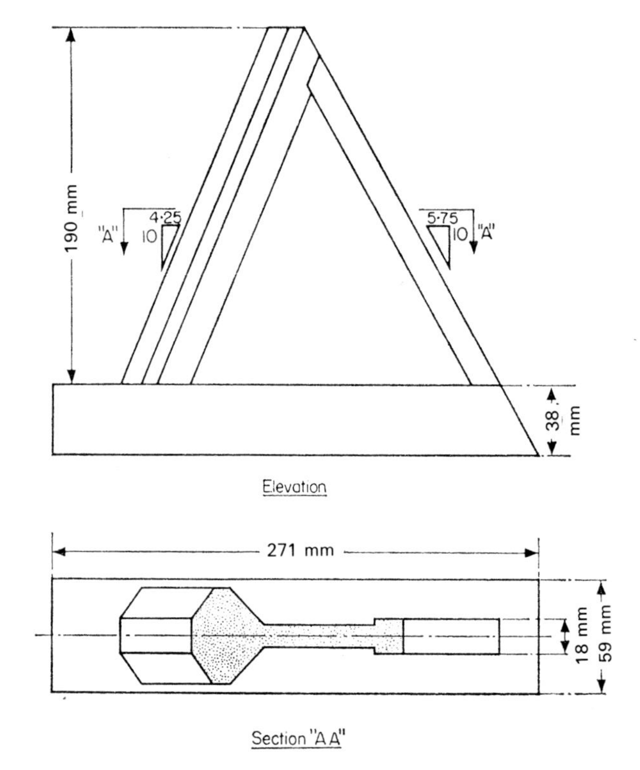

Fig. 12.3. Model of diamond head buttress unit for Errochty dam. Scale of model 25 mm = 6000 mm.

feasible method for this investigation. Gravitational loading was simulated by subjecting the model to an acceleration field equivalent to 45 g in a large centrifuge. Hydrostatic loading was approximated, separately, by a bank of 24 small hydraulic rams acting on the faces of the buttress; the piston diameters were varied so as to produce a stepped loading and all the rams were supplied from a common master cylinder. In a subsequent investigation of the stresses in a dam, the author was able to simulate hydrostatic and gravitational loadings simultaneously by applying fluid pressure to the face of the model dam in the centrifuge. The loading fluid was methyl stearate, a substance which melts at 35°C and which was thus fluid through the softening range of the plastic but solid at room temperature, so that it did not fall out of the loading box when the centrifuge was running up to speed.

Torsional Stiffness of a Reinforced Concrete Tower [4]

This problem was investigated by Sparkes and Chapman in connection with the design of the C.I.S. Office Building in Manchester. The nature and purpose of the investigation may be summarised as follows:

1. Purpose: information for design
2. Extent of investigation: to determine the torsional stiffness and shear centre of the structure and to determine strains due to torsional and flexural loading at important points in order to check the proposed reinforcement
3. Laboratory investigation
4. Lift shaft and stairway structure for a high building
5. Prototype—reinforced concrete
6. Elastic analysis
7. Arbitrary stress scale to suit method of analysis
8. Twisting and bending moments on the tower arising from wind loads on the tower and on adjoining steel structures
9. Short term tests
10. Test primarily concerned with determination of torsional stiffness
11. No special limitations regarding time and cost

mentioned by investigators but need for reasonable economy and expedition may be assumed.

The essential problem in this investigation was the determination of the torsional stiffness of the tower structure, the section of which was neither completely open nor completely closed. The outer walls form an open channel section, whilst the inner walls are joined by three vertical open webbed girders;

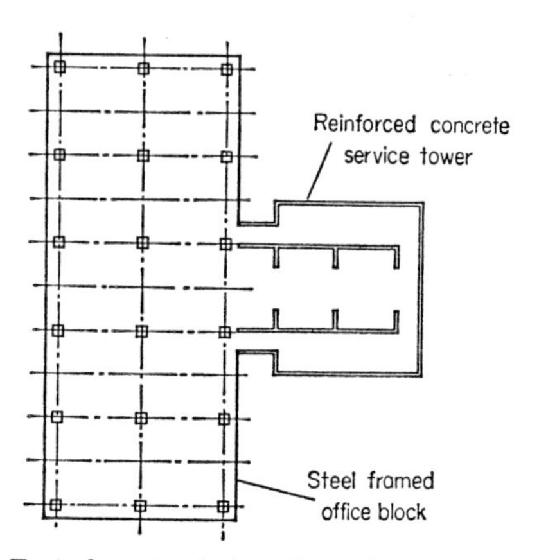

Fig. 12.4. Typical sectional plan of one floor of C.I.S. Building, Manchester.

the outer and inner systems are joined by horizontal slabs and beams. The calculation of the effective torsional stiffness of a complex structure of this kind is obviously extremely difficult, but if it could be established that the tower was sufficiently stiff to resist side-sway forces and wind loads, important economies would be possible in the design.

Sparkes and Chapman selected Perspex for the construction of the model for the following reasons: (i) its favourable machining properties, (ii) the fact that it is easily joined using an acrylic resin cement (Tensol 7), (iii) it does not creep excessively and (iv) it can be obtained from stock in a range of thicknesses. The scale of 1:60 was selected after weighing the problem of fabrication and measurement against the cost of materials. The model was loaded by dead weight through cables passing over pulleys. Deflections were measured at six levels by dial gauges and strains at about 100 stations by electrical resistance strain gauges.

Vibration of Grid Frameworks

Outline of Problem:

1. Purpose: Research
2. Extent of investigation: determination of critical frequencies and modes of vibration of structure
3. Laboratory investigation
4. Interconnected beam systems
5. No specific prototype
6. Elastic behaviour
7. Vibrational loading
8. Short term readings
9. C.R. oscilloscope, ultraviolet recording galvanometer, inductance accelerometers, etc., available
10. Limited funds available for additional equipment.

As noted under (2) above, the object of this investigation was to determine modes and frequencies of vibration of interconnected beam structures. This information was required for the verification of a theoretical solution of the problem. It was considered convenient to construct model grids of 900 mm span using Perspex members. The reasons for using this material being similar to those stated by Sparkes and Chapman in connection with their study of the C.I.S. tower. In order to determine the natural frequencies of a frame, the structure was caused to vibrate by an electromagnetic exciter adapted from a 30-watt loudspeaker unit. This was fed from a signal generator through a suitable power amplifier. The accelerations at any selected point in the frame were picked up by an inductance accelerometer, the output from which was displayed on a cathode ray oscilloscope. The critical frequencies could thus be determined by searching the frequency range for maximum displacements of the structure. Although displacements could be determined from the acceleration measurement, it was not possible to ascertain the mode of vibration of the frame from these observations. This difficulty was resolved by taking simultaneous strain gauge recordings on the centre transversal of the frame so that the relative positions of the longitudinals could be determined at any instant. For this part of the work, the accelerometers and strain gauge signals were recorded by means of a high speed ultraviolet recording galvanometer. Although the number of channels available was limited, it was possible to determine the complete behaviour of the structure by these means.

Fig. 12.5. Grid framework; dynamic analysis.

CONCLUSION

These examples could be continued indefinitely, but may be sufficient to illustrate the selection of method of analysis, materials, scales, loading gear and so on in typical problems in experimental stress analysis. It is probably safe to say that electrical resistance strain gauge analysis is the most widely used method, particularly in laboratory investigations. For measurements on full size structures, which usually extend over a prolonged period, there is some preference, at least in the United Kingdom, for vibrating wire gauges which are not readily affected by moisture and temperature changes. Simple mechanical gauges are often useful in this kind of work if the members permit reasonably large gauge lengths and if the points of measurement are accessible.

Photo-elasticity has the advantage of giving a complete picture of the stress field under investigation. Further, it makes use of comparatively small models, which in turn require light and comparatively inexpensive loading gear. The method is particularly useful for the investigation of detailed stress distribution in elements of complicated geometry. The photo-elastic coating method is likely to be of value in examining stress concentrations in pressure vessels and the like and for check tests on full sized structures of actual constructional materials.

A number of methods not mentioned in the examples discussed in this chapter are rather specialised and thus their application is obvious. For example, the moiré fringe method was specifically devised for the investigation of the stresses in plate structures and the Begg's deformeter method and its later developments are useful only for the analysis of statically indeterminate structures of various types.

The decision to use an analogue method depends very much upon how closely the given loading and boundary condition can be reproduced on the analogue, how quickly the problem can be set up and how often the analogue is likely to be used. The point of this last remark is that there would be little advantage in setting up a complicated analogue for one particular investigation. If the problem is a fairly routine one, likely to be encountered frequently, and for which a convenient analogue is available, there is much to be said for having the equipment at hand. The solution of a given problem will then be very rapidly obtained. An example of this kind might be the determination of the torsional characteristics of a non-circular section by the conducting paper method. Similarly, the availability of a network analogue for the solution of frame problems may be well worth while in a large design office or in a research department, but in this situation analogue methods have been almost completely displaced by the electronic digital computer.

To make the correct selection of method, requires a good knowledge of the range of available techniques and a certain determination to approach each problem with an open mind: much misplaced ingenuity has been expended on unsuitable applications by enthusiasts of one method or another. It is immensely helpful to make a systematic examination of reported investigations and the references given at the end of this chapter will provide the student with some further examples of experimental study of the stresses in structures.

BIBLIOGRAPHY

1. A. W. HENDRY, The stress distribution in welded steel portal frame knees, *Struct. Eng.*, **25**, 101–141 (1947).
2. A. W. HENDRY and D. A. TURABI, Theoretical and measured stresses in a welded steel dock gate, *Proc. Inst. C.E.*, 537–48 (1961).
3. A. W. HENDRY, Photoelastic experiments on the stress distribution in a diamond head buttress dam, *Proc. Inst. C.E.*, Part I, 370–96 (May 1954).
4. S. R. SPARKES and J. C. CHAPMAN, Model methods, with particular reference to three recent applications in the fields of steel, composite and concrete construction, *Struct. Eng.* **39** (3), 85–99 (1961).
5. M. ROCHA, *Model study of structures in Portugal*, Pub. 84, Lab. Nac. de Eng. Civ., Lisbon, 1956. (English translation available as TT-970, Nat. Res. Council Canada Div. of Bldg. Res., Ottawa, 1961).
6. G. OBERTI, Large scale model testing of structures outside the elastic limit, *RILEM Bull.*, **7**, 40–58 (1960).
7. ——, Proc. RILEM Symposium on models of structures, Madrid, 1959, *RILEM Bulls.*, 7–11 (1960–61).

8. G. OBERTI, *Prelim. and Final Vols. Conf. on Correlation between calculated and observed stresses in structures.* Inst. C.E., London, 1955.
9. ——, *Proc. Symposium on the Observation of Structures, Lisbon,* 1955, 2 vols. Lab. Nac. de Eng. Civ., Lisbon, 1955.
10. M. ROCHA, Experimental dimensioning of structures, *Ann. Inst. Tech. Bât. Trav. Publics.* (*N.S.*), 235 (1952).
11. L. CHITTY and A. J. S. PIPPARD, The determination of the stresses in an arch dam from a rubber model, *Proc. Inst. C.E.*, Part 1, **5** (3), 259–275 (1956).
12. D. N. ALLEN *et al.*, The experimental and mathematical analysis of arch dams with special reference to Dokan, *Proc. Inst. C.E.*, Part 1, **5**, (3), 198–258 (1956).
13. M. ROCHA and J. L. SERAFIM, *The Problem of the Safety of Arch Dams—Rupture Studies on Models.* Minis. Obris. Publicas, Lab. Eng. Civ., Lisbon, Pub. 142, 1960.
14. R. KUHN, Experiences in strain measurements in concrete bodies of large dimensions, *ZVDI*, **99** (17), 751–63 (1957).
15. R. HILTSCHER and R. K. MULLER, Analysis of the concrete reinforcement of steel construction with the aid of photoelasticity, *Beton u. Stahlbetonbau*, **54** (11), 263–271 (1959).
16. J. L. WILBY and L. B. GRIZZUK, Experiments with thin shell structural models, *J. Am. Conc. Inst.*, **32** (4), 413–432 (1960).
17. R. E. ROWE, *Tests on four types of Hyperbolic Shell*, Proc. Symposium on Shell Research, Delft, pp. 16–35. J. Wiley, New York, 1961. (See also other similar papers in this publication.)
18. R. M. FINCH and A. GOLDSTEIN, Clifton Bridge, Nottingham; initial design studies and model test, *Proc. Inst. C.E.*, **12**, 289–316 (1959).
19. A. LITTLE, The distribution of a load in a box-section bridge from tests on a Xylonite model, *Mag. Conc. Res.*, **6** (18), 121–132 (1954).
20. D. H. PLETTA *et al.*, Tests of rigid-frame bridge model to ultimate load, *Proc. Am. Conc. Inst.*, **58** (2), 223–42 (1961).
21. J. M. RUZEK *et al.*, Welded portal frames tested to collapse, *Proc. Soc. Exp. Stress Anal.*, **11** (1), 159–180 (1953).
22. R. S. ROWE and S. SHORE, Model arches in the flexible range; testing technique, *Proc. Soc. Exp. Stress Anal.*, **9** (2), 31–42 (1952).
23. V. KOLOUSEK, Vibrations of bridges with continuous main girders, *Pub. Int. Assn. Bridge Struct. Eng.*, **19**, 111–137 (1959).
24. E. ZELLERER and H. THIEL, Force field in a wall beam on two supports with three door openings in the case of unsymmetric concentrated forces, *Bautechnik*, **36** (3), 84–95 (1959).
25. A. W. HENDRY and S. SAAD, Gravitational stresses in deep beams, *Struct. Engineer*, **39** (6), 185–94 (1961).

INDEX